MAD LIKE TESLA
Tyler Hamilton

MAD
LIKE
TESLA

Underdog Inventors
and their **Relentless Pursuit**
of Clean Energy

TYLER HAMILTON

Published by ECW Press
2120 Queen Street East, Suite 200, Toronto, Ontario, Canada M4E 1E2
416-694-3348 / info@ecwpress.com

LIBRARY AND ARCHIVES CANADA CATALOGUING IN PUBLICATION

Hamilton, Tyler J.
Mad like Tesla : underdog inventors and their relentless
pursuit of clean energy / Tyler Hamilton.

ISBN 978-1-77041-008-4
ALSO ISSUED AS: 978-1-77090-074-5 (PDF) ; 978-1-77090-073-8 (EPUB)

1. Renewable energy sources—Popular works. 2. Energy
development—Popular works. 3. Inventors. I. Title.

TJ808.H36 2011   333.79'4   C2011-902853-0

Editor for the press: Crissy Boylan
Cover and text design: David Gee
Typesetting: Rachel Ironstone
Production: Troy Cunningham
ECW PRESS   Printing: Edwards Brothers
ecwpress.com

Interior images by Tyler Hamilton, except p. 74/75, illustrations by Tania Craan based on an illustration by Brian Kaas for Sierra magazine; p. 120 PAX Scientific; and p. 153/154 illustrations by Tania Craan based on illustrations by Algenol Biofuels.

The publication of Mad Like Tesla has been generously supported by the Canada Council for the Arts, which last year invested $20.1 million in writing and publishing throughout Canada, by the Ontario Arts Council, by the Government of Ontario through Ontario Book Publishing Tax Credit, by the OMDC Book Fund, an initiative of the Ontario Media Development Corporation, and by the Government of Canada through the Canada Book Fund.

 Canada Council   Conseil des Arts      ONTARIO ARTS COUNCIL
for the Arts   du Canada   Canada   CONSEIL DES ARTS DE L'ONTARIO

PRINTED AND BOUND IN THE UNITED STATES

for my bundles of joy,
Claire and Ruby

CONTENTS

INTRODUCTION
Cheering the Lone Runner

It will never work. It can never be done. It is impossible. It will never be accepted.

How often, throughout modern history, have those words been spoken? There are dozens of classic examples. Physicist and engineer Lord Kelvin, president of the British Royal Society, famously said back in the late 1800s that "heavier-than-air flying machines are impossible" and "x-rays will prove to be a hoax." Albert Einstein said in 1932 that he couldn't see nuclear energy ever being obtainable. Tunis Craven, in his role as commissioner of the U.S. Federal Communications Commission in 1961, spoke a few years prematurely when he dismissed the future prospect of communications satellites. "There is practically no chance communications space satellites will be used to provide better telephone, telegraph, television, or radio service inside the United States," pronounced Craven, only to be proven wrong when the satellite Syncom 3 transmitted television signals from Japan to the United States, giving Americans live foreign coverage of the 1964 Summer Olympics. Many theoretical physicists during the late 1960s and '70s said nuclear magnetic resonance technology, first discovered in the 1930s, could not be adapted to detect cancers and other diseases in the human body. A New York physician named Raymond Damadian ignored the naysayers and built his own body scanner. In 1977 he successfully performed the first full-body magnetic resonance imaging (MRI) exam on a human. Receiving an award from the Massachusetts Institute of Technology in 2001, Damadian said criticism and skepticism come with the territory of invention and innovation: "The bolder the initiative, the harsher the criticism."[1] A gracious response, but I rather like the comment by *Time* magazine writer Lev Grossman, "There's nothing like the passage of time to make the world's smartest people look like complete idiots."[2]

In the area of energy technology and systems, it's arguable that no one has been doubted, underestimated, or challenged more than Serbian-American engineer Nikola Tesla (1856–1943), and perhaps no innovator has proven so many people wrong over the past 100 years. His best-known invention was the alternating current (AC) induction motor, patented in 1888, which became crucial to the subsequent development of high-voltage AC power systems that could distribute electricity over long distances. Direct current (DC) systems of the day, by comparison, were limited because they produced electricity that had to be consumed within a couple of kilometers of where it was generated. Tesla appreciated early in his life the benefits of AC and the shortfalls of DC, and he first began formulating his AC motor designs as a student at the Austrian Polytechnic School in Graz. One day, after sharing his thoughts with a professor he greatly respected, the young Tesla, in his early 20s, was promptly rebuffed in front of his peers. "Mr. Tesla may accomplish great things," his professor quipped, "but he will never do this . . . It's a perpetual motion machine, an impossible idea."[3] Of course, what Tesla described had nothing to do with perpetual motion. Yet his professor, unable to grasp the concept, did not hesitate to lump it into the category reserved for loony inventions.

A few years later, while walking in a park late one afternoon with a friend, Tesla said a clear design for his motor shot into his head "like a flash of lightning." He found a stick and began drawing diagrams in the dirt. Six years later, he formally revealed the design to the American Institute of Electrical Engineers, though the battle to have it more widely accepted was just beginning.[4] As he forged on and turned his idea into a working prototype, he faced resistance from American inventor Thomas Edison, who saw all inventions around alternating current as a threat to the low-voltage, direct-current systems he was developing and from which he was collecting handsome royalties. "Fooling around with alternating current is just a waste of time," said Edison in 1889. "Nobody will use it, ever." To discourage its use, Edison declared high-voltage AC systems unsafe and lobbied the U.S. government to ban the technology on those grounds.[5] He drove the message home by funding a vicious public relations campaign that involved electrocuting dogs and other animals — including an elephant — with AC current. His campaign didn't work, of course, and AC power generation and transmission systems based on Tesla's designs eventually came to dominance around the world. It wasn't an impossible waste of time after all.

I first became fascinated with the life and work of Nikola Tesla 10 years ago while researching a story marking the 100th anniversary of Guglielmo Marconi's first transatlantic wireless communication. On December 12, 1901, the letter "S" was transmitted in Morse code from a wireless transmission station in Cornwall, United Kingdom, to Signal Hill in St. John's, a city in the Canadian province of Newfoundland and Labrador. The 3,500-kilometer transmission across the Atlantic Ocean was hailed as a great moment in science and eventually led to Marconi being branded as "The Father of Radio." It's a nice story, except for the fact that Tesla was the true father of radio. Tesla filed his first radio patents in 1896, five years before Marconi, and when Marconi did file to the U.S. patent office, it initially rejected his application for being too similar to those submitted by Tesla. It was only when the U.S. Supreme Court ruled on the patent issue nearly 50 years later, in 1943 just a few months after Tesla's death, that the matter was settled for good: Tesla invented radio.

The more I dug into the life of Tesla, the more I realized how much his inventions and vision influenced the 20th century and continue to do so in our current century. In 1898 he demonstrated a remote-controlled robotic boat that heralded the beginning of remote-controlled electronics and multichannel broadcasting. Tesla called it "teleautomation" and referred to his boat as "the first of a race of robots — mechanical men which will do the laborious work of the human race." Learning this gave new meaning to the remote-controlled, battery-powered Spiderman helicopter I fly with my kids in the basement, though the larger influence is obvious — everything from the Mars Rover to unmanned drone planes to cruise missiles.[6] To those watching the demonstration in 1898, Tesla's invention seemed powered by pure magic. "Skeptics had him pull the lid to prove there wasn't a midget operating inside," according to documentarian Robert Uth.[7] But there was little financial interest in the idea at the time, and Tesla explained such reluctance to invest — and even hostility to the concept — when he requested development funds from American millionaire John Jacob Astor IV. "It is for a reason that I am often and viciously attacked," he wrote in a letter to Astor, "because my intentions threaten a number of established industries."[8] Astor, who went down with the *Titanic* in 1912 at the age of 47, would only live long enough to get a taste of how Tesla's robot race would evolve.

There seems to be no end to Tesla's accomplishments. He is credited in hindsight for perfecting neon and fluorescent lighting, for the earliest work in wireless power transmission, for taking the world's first x-ray photographs, and for proposing the basic principles of radar technology two decades before its "official" invention.[9] Even his ability to create "lightning balls" in the laboratory has inspired research into plasma physics and nuclear fusion. An inventor and visionary, Tesla was also an environmentalist at heart who understood the long-term implications of burning fossil fuels, the supply of which would eventually run out. Non-renewable resources such as coal and oil should be conserved, he argued, and solar energy — either acquired directly or indirectly by capturing energy in the wind — should be fully exploited. At age 75, he published designs of two renewable energy extraction methods in the December 1931 issue of *Everyday Science and Mechanics*. One was an early design of a geothermal power plant of the sort built today to extract heat from deep underground for clean electricity production. The second was a way to generate electricity from the temperature differential in ocean waters.

Essentially, a heat exchanger would extract heat out of the warmer upper ocean layers and create steam from a working fluid with a low boiling temperature, such as ammonia. The steam would drive a turbine that generates electricity. Cold water from deeper layers would then be used to condense the ammonia back into fluid, at which point the cycle would be repeated. It seemed an unpractical and somewhat wonky concept at the time, and Tesla himself struggled to make it efficient enough to be worthwhile. But the idea, today called ocean thermal energy conversion (OTEC), lives on. Military contractor Lockheed Martin has been working away on the technology since the 1970s and is now constructing a 10-megawatt OTEC pilot plant off the coast of Hawaii that could be in operation by 2013. It would generate enough power for 10,000 homes. "I dream of thousands of floating OTEC ships roaming the seas of the world, providing an inexhaustible supply of clean energy and fuel and water for all people of the world," explained an enthusiastic Ted Johnson, director of alternative energy development at Lockheed.[10] It's a comment Tesla might have made 100 years ago.

It's no wonder Tesla is regarded today as one of the greatest inventors and thinkers of the 20th century. But the label only took hold many years after his death. And it didn't come without a fight. In many respects, he was an underdog his entire career, despite his lifelong creativity and brilliance. Born July 10, 1856, in the Croatian village of Smiljan, the young Nikola was a sickly and quirky child who developed many phobias and obsessions that were carried into adulthood. When he walked, he couldn't help but count his steps. When served meals, he was compelled to calculate in his mind the cubic contents of cups and bowls. He also had a problem touching people's hair and, generally, had a phobia of germs. The sight of earrings dangling from a woman's earlobes disgusted him, especially if they were pearls. This developed much later in life into other obsessive-compulsive behaviors, most notably a fixation on the number three. He often walked around the block three times before entering a building, or he would demand that certain items be sent to him in threes. Such eccentric behavior may explain why Tesla never married or engaged in romantic relationships during his life. Poor and feeble in his final years, his only known "love" was a white pigeon that would regularly visit his apartment. It was during these final years that Tesla also spoke openly about a future that most people could not imagine, let alone believe.

According to the documentarian Uth, "Newspaper reporters, looking for a laugh, would attend to listen to the crazy old man's outlandish predictions of wireless telephones, communication with other life forms in the cosmos, beam weapons that could shoot down airplanes and missiles, and many more science fiction concepts that are now becoming reality." Instead of being hailed as a genius and brilliant inventor, Tesla was increasingly thought of as a "mad scientist" in the minds of the general public and in popular culture. In 1941, the first Superman cartoon movies, for example, had the Man of Steel fighting a mad scientist named Tesla and his city-destroying "electrothanasia ray," representing Tesla's real-life work on a particle beam that could be used as a weapon of war. Perhaps Tesla seeded such perceptions of madness as early as 1901 when he claimed he had been in communication with Mars, implying some sort of back-and-forth chatter with extraterrestrials. Conspiracy theorists suggest such communications informed Tesla's reported early awareness of rising carbon dioxide levels in Earth's atmosphere and how it was causing global warming.[11] Needless to say, Tesla was just as often a target of ridicule as he was of praise. "He was a really weird dude," Amory Lovins, chief scientist and co-founder of the Rocky Mountain Institute in Colorado, told me one day while I was researching this book. "He probably would have been in a mental institution today."

One wonders how many brilliant minds are undiscovered, locked away or medicated, and to what extent our aversion to "weird dudes" puts a cap on innovation. Another question is whether Tesla's strange ways contributed in a way to his genius, or whether it was his genius that led to his strange ways. I keep coming back to a line in the Francis Ford Coppola movie *Rumble Fish*, in which Dennis Hopper's character answers a not-so-simple question from his son Rusty James (Matt Dillon): was mom crazy? "Every now and then a person comes along that has a different view of the world than does the usual person," says Hopper's character. "It doesn't make them crazy. I mean, an acute perception, man — that doesn't make you crazy. However, sometimes it can drive you crazy." What we know is that Tesla's acute perception of the world around him informed a kind of thinking that most others could not easily grasp or appreciate. To be misunderstood can make one appear mad in the eyes of others. But being misunderstood, over time, is just as likely to drive one mad.

A voracious reader, a lover of poetry, and a man blessed with a photographic memory, Tesla somehow managed to balance his eccentric ways with social acceptance for most of his life. Mark Twain enjoyed hanging around him, and the two became close friends. The social elite of New York City, where Tesla lived, loved to rub elbows with the strange-but-fascinating inventor who had a lovely accent and slim build. But, ultimately, the Serbian-American engineer was a loner both in how he lived and in his head. He kept ideas locked up inside his mind and, to the frustration of many, did not collaborate well with other engineers and scientists. His ideas and achievements were his and his alone, as was his struggle to have those innovations accepted by society. This isolation, and the fact that he has proven so many skeptics wrong over the years, has endeared Tesla to many inventors and entrepreneurs who identify with his struggle. "The example set by Tesla has always been particularly inspiring to the lone runner," wrote Margaret Cheney, author of *Tesla: Man Out of Time*.

Indeed, one of the inventor's 21st century namesakes is electric car maker Tesla Motors, which relies on a modern version of Nikola Tesla's "polyphase" AC motor design. The company's Tesla Roadster is a $109,000 battery-powered sports car with a 230-mile range and acceleration that can leave a Porsche in its dust. Its early critics considered it a flash in the pan — a shiny, fast toy for the uber-rich that won't last beyond its initial buzz. Many still view the company this way. As a pioneer of electric cars, and the first automaker born out of Silicon Valley, Tesla Motors is very much a lone runner that's used to being told "it can never been done" and "it will never be accepted." Yet the company, like the inventor, has not been discouraged by the naysayers. This is made clear on its corporate website: "The critics said it couldn't be done, yet we are here, taking nothing for granted. We challenge custom and question tradition. Our drivers benefit from it." Since Tesla Motors emerged on the scene, most major automakers, including those that declared electric cars a short-lived fad, have announced plans for their own all-electric or plug-in hybrid vehicles. The global transition to electrified transportation is building momentum, and it's now viewed by many as almost inevitable.

This book explores some of the other "lone runners" out there who I have identified during my six years as an energy reporter and clean technology columnist for the *Toronto Star*, Canada's largest daily newspaper, and as a frequent contributor to MIT's *Technology Review*. I titled it *Mad Like Tesla* because, in my observation, the companies and individuals profiled here have reason to identify with Nikola Tesla, the man. They are considered — or have been considered — crazy because of the perceived impossibility or unacceptability of what they're attempting to do, and yet they forge ahead in an inhospitable marketplace driven largely by a desire to do right and a conviction that they are right. In this sense, they are mad like Tesla — not because women's earrings make them vomit or they claim to have communicated with Martians, but because of how easily society dismisses their potentially game-changing efforts and because of the barriers they face along their journeys. The barriers are many: scientific groupthink, bad timing, entrenched corporate interests, misplaced public fear, gaps in available technology, high cost, resource scarcity, personality clashes, lack of financing, resistance to change, complacency, competitive rivalry, misguided policy, lack of vision, and general ignorance — to name just a few. Many of these barriers will be discussed in the chapters that follow. "It's a surprise some people ever start," industry watcher Rick Whittaker once told me. He's the chief technology and investment officer at Sustainable Development Technology Canada (SDTC), a federal granting agency created to support clean technology demonstration projects. Hundreds of funding applications cross Whittaker's desk each year. Many deserving ideas slip through the cracks. Those persistent or lucky enough to get funding are only at the beginning of a very long road. The casualty rate is high.

So why did I write this book? Humanity is on an unsustainable path, and changing course will require a dramatic rethink of how we obtain and use energy. We need to be more open minded. We need to take more chances. The individuals and companies profiled in the chapters that follow play a crucial role in our energy future, even if they fail. That's because, like Tesla, they still succeed by leading, by taking risks, by pursuing great leaps, and by keeping open minds when others remain so closed. They stand in contrast to monolithic corporations with disciplined management cultures and an aversion to disruptive technology. As U.S. inventor Dean Kamen, creator of the Segway scooter, once said, "Good management tries to eliminate surprise, therefore good management eliminates innovation." Fact is we need the lone runner, be it a passionate individual or an aggressive startup that doesn't stop at "no" and isn't satisfied with taking baby steps. But who are these people? What drives them? What goes through their minds? Though we see the sensational headlines that briefly shine a light on their unusual technologies — the so-called 15 minutes of fame — too often these innovators' stories and their contributions to the world fade into the background. The real learning comes out of the journey behind the headline.

But let's be clear: the nimble and creative David isn't guaranteed to come up with a better concept or technology than a slow and stifled Goliath. Even Tesla had his duds, as author Judy Wearing illustrates so well in her book, *Edison's Concrete Piano*. Tesla's "earthquake machine" wasn't of much value, and his cosmic theories aimed at refuting Einstein's theory of relativity were well off the mark. I also don't want to suggest that simply proving a ground-breaking idea both technologically possible and superior is all it takes to earn rapid acceptance, as writer Vaclav Smil discusses in his book *Energy Myths and Realities*. "Wishful thinking, pioneering enthusiasm, and belief in the efficacy of seemingly superior solutions are not enough to change the fundamental nature of gradually unfolding energy transitions, be they shifts to new fuels, to new modes of electricity generation, or to new prime movers," wrote Smil. His point is well taken. Pieces of an energy system aren't iPod-like gadgets with a six-month shelf life. They are part of a massive infrastructure that has come together over several decades at a cost of hundreds of billions of dollars. Meaningful change on a global scale will take many decades more, so we'll need to temper passion with patience.

Energy expert David Fridley, a scientist at Lawrence Berkeley National Laboratory in California, writes that too many of us underestimate the grueling path to commercialization, even for truly breakthrough energy technologies. "Processes need to be perfected and optimized, patents developed, demonstration tests performed, pilot plants built and evaluated, environmental impacts assessed, and engineering, design, siting, financing, economic and other studies are undertaken. In other words, technologies that are proved feasible on the bench-top today will likely have little impact until the 2030s."[12] It's also not a sure thing that a new game-changing technology can be scaled up enough to have the kind of global impact that's expected. Maybe it relies on a rare-earth metal that's in very short supply or is limited by geography or needs to consume huge volumes of fresh water at a time when water is increasingly scarce. These variables must be carefully weighed and considered.

With all of that duly noted, in Chapter 1 you'll meet Michel Laberge, a Quebec-born engineer who worked for a high-tech commercial printing company until he turned 40 and had what he considers a mid-life crisis. Most men would buy a Porsche and start dyeing their graying hair. Not Laberge. He decided to dedicate his life to building a nuclear fusion power reactor on the cheap. His Vancouver-based company, General Fusion, is trying to do with tens of millions of dollars what government-led projects in Europe and North America are struggling to do with tens of billions. He understood the odds were against him when he started, but ask him today and he'll tell you it's a 50–50 bet and the odds are getting better every day. If he pulls it off, clean and cheap nuclear power without the toxic waste just might be a reality in our lifetime.

Chapter 2 introduces you to a California company that wants to go where no power plant has gone before — space. Solar photovoltaic technology has been used for decades to power satellites, but Gary Spirnak wants to take an idea first proposed by science fiction writer Isaac Asimov in 1941 and — in the words of next-gen *Star Trek* captain Jean-Luc Picard — make it so. A square kilometer solar collector would be launched into orbit about 36,000 kilometers above the surface of the Earth, clear of clouds and facing the sun 24 hours a day. The energy collected would be beamed by microwave down to a massive receiving station in the middle of a desert, converted into electricity, and put on the power grid. Sounds crazy, I'll admit. Spirnak, founder of Solaren Corporation, admits it's an enormous endeavor that invites ridicule, but, as an engineer and veteran of the U.S. space industry, he is convinced it can be done with current technology and that the electricity produced will be competitively priced.

As Solaren works toward harnessing solar energy from space, Louis Michaud is caught up in the idea of tornado power. The retired refinery engineer from a part of Ontario, Canada, known as "chemical valley" has spent much of his adult life studying how temperature differentials between the ground and the troposphere lead to the creation of tornados, dust devils, and, over the ocean, water spouts. These naturally occurring vortices contain immense amounts of energy. In Chapter 3 you'll find out how Michaud wants to harness that energy by creating man-made tornados that would "survive" using waste heat from industry or the warm ocean waters of the tropics. He calls his invention the atmospheric vortex engine, and while a small-scale version proves the concept works — including a prototype built in Michaud's garage — taking it to the next level has been his greatest challenge.

Finding new ways to generate clean electricity is only part of the challenge the world faces. We also need to use less energy to accomplish the work we desire, and some inventors are taking their cues from nature. Like Australian entrepreneur Jay Harman, who observed early in his life that certain shapes and patterns found naturally in the world — the swirling of ocean kelp, the cross section of nautilus seashells, the shape and movement of tornados — exist for a reason. Simply put, they're more efficient at managing the flow of fluid or air. In Chapter 4 you'll learn about Harman's company, PAX Scientific, and his frustrating attempts to convince the captains of industry, and investors, that everything from fans and turbines to pumps and mixers can be made dramatically more efficient. You'll also be introduced to the emerging field of biomimicry, as well as a number of other inventors who, like Harman, are borrowing from Mother Nature's cookbook.

The sustainability imperative also has us targeting humanity's addiction to oil, which according to the International Energy Agency represents roughly 30 percent of the world's primary fuel mix, most of it guzzled by the transportation sector.[13] Yes, electric vehicles are on the way, but many will be plug-in hybrids that still burn gasoline, and their introduction into use will start slowly. Also, airplanes can't fly on battery power. There will remain a need for green fuel alternatives, such as ethanol, a multi-purpose substance which can also replace the use of environmentally harmful petrochemicals in the making of plastics and other materials. In Chapter 5 we visit a Florida startup called Algenol that has come up with a clever way to grow ethanol-producing algae on an industrial scale. There is no shortage today of algae-to-fuel ventures, but most grow algae so they can be harvested — that is, killed — and squeezed of their natural oils. Algenol treats algae more like dairy cows. It grows them on a steady diet of sunlight and carbon dioxide and then milks them. Paul Woods came up with the unusual idea during the 1980s when he was just 22, but nobody then believed it could be done. He stayed at it, self-funding the project over the years with help from friends and family. Today Algenol is one of the most innovative biofuel companies on the market, and its approach to producing green fuel from algae could end up setting a standard for the industry. But Woods' journey, as you'll read, has in many ways just begun.

Better technologies for storing electricity could also significantly reduce humanity's dependence on oil, not to mention coal and natural gas. That's the solution Dick Weir is targeting. The co-founder of Texas startup EEStor, explored in Chapter 6, is working on what he claims will be the ultimate energy storage device, capable of keeping an electric vehicle charged during a 500-kilometer drive or storing huge amounts of energy from the sun and wind so it can be dispatched as needed. Laptops, power tools, and other gadgets would be able to run for weeks on a single charge. A game-changing development? Yes. But is it viable? Many engineers and scientists have their doubts. EEStor and its crusty founder are secretive, stay clear of the media, and don't like being rushed. All of this has shrouded the company in mystery, attracting a mix of fans and critics and the kind of Internet gossip typically associated with celebrities like Paris Hilton. Energy storage is widely recognized as the nut that needs to be cracked if we are to see renewable energy become a serious threat to fossil fuels. EEStor intends to crack it, despite the army of skeptics who seemingly wish it to fail.

Skepticism is integral to science and invention, as there is no shortage of folks who claim to have discovered some variation on perpetual motion. Can a law of physics be broken? Do we know all there is to know? These are divisive questions in the scientific community, and debate can take on a religious tone. This is the territory we enter in Chapter 7, where you'll be introduced to Thane Heins and his controversial Perepiteia Generator. Does it put out more power than it consumes, as Heins claims? It's impossible, according to the laws of physics, yet in laboratory demonstrations the behavior of his machine has stumped believers and skeptics alike. Heins is an unlikely inventor, but, as Tesla once said of himself, he has the "boldness of ignorance." His endeavor may turn out a fool's voyage, but with fans like legendary rocker Neil Young in his corner — seriously, Neil Young — the memories will certainly be precious.

Finally, I end the book by posing this question: can tech-boom billionaires and the pioneering spirit of Silicon Valley accelerate our transition to a low-carbon economy? There's a belief out there that investors and entrepreneurs who got rich from huge bets in computing, the Internet, and telecommunications — and transformed each industry seemingly overnight — can do the same with energy; that their growing economic influence can speed up the acceptance and adoption of clean energy technologies that might otherwise be ignored by big industry. Acknowledging that their participation can only help, I wonder if the technology world's top venture capitalists, wealthiest entrepreneurs, and most seasoned executives have overestimated their own influence and underestimated the energy challenge. Ultimately, it may be "Black Swans" that determine our energy destiny.

The stories you are about to read concern science and technology with a splash of business and economics and a dash of recent history. To stretch the cooking analogy, it's all served up on a platter of human interest with a side order of Tesla. The occasional link back to Nikola Tesla reminds us that the obstacles to energy innovations that existed 100 years ago remain in the 21st century. At the same time, it's important to recognize that the economic and environmental drivers of innovation are arguably very different today; there was no sustainability imperative driving invention in the early 1900s.

In Tesla's time there was certainly little, if any, awareness of climate change and the risks of using our atmosphere as a dumping ground for carbon dioxide and other greenhouse gases. We know now that dependence on fossil fuels on an increasingly crowded planet is not sustainable, and that human-caused climate change is contributing to extreme weather, record high temperatures, the melting of the Arctic and glaciers, and the mass extinction of species on land and in water. A July 2010 report from the U.S. National Oceanic and Atmospheric Administration made this disturbingly clear. Scientists collected data from 160 research groups in 48 countries to get a comprehensive picture of how rising greenhouse-gas concentrations are affecting the planet. Going back in some cases more than 100 years, they identified an unmistakable trend. Air temperature over land and sea is going up, ocean surface temperature is rising, sea levels are rising, humidity is getting worse, and the troposphere continues to warm. Meanwhile, we're seeing clear evidence of declining Arctic sea ice, retreating glaciers, and decreasing spring snow cover in the Northern Hemisphere. The years 2000 to 2009 were also declared the hottest decade on record.[14] "Climate change is occurring, is caused largely by human activities, and poses significant risks for — and in many cases, is already affecting — a broad range of human and natural systems," warned another report, released just two months earlier, from the prestigious U.S. National Academy of Sciences.[15]

I will never forget a private meeting I attended in Toronto on May 8, 2009, with British scientist James Lovelock, whose Gaia theory explains the Earth's biosphere as a self-regulating entity quite capable of adapting to climate change. Humans, however, will be, in his view, a casualty of that adaptation. Commenting on the impact of climate change over the next few decades,

Lovelock, two months shy of his 90th birthday, painted a shockingly grim picture. "Anything that overgrows its resources gets smacked back down," he told us. "I foresee a loss of as much as 80 or 90 percent of the people on Earth by the end of the century. It's a distinct possibility, and I don't think there is much we can do to stop it. You have to make sure those who remain will be able to survive it." I still recall looking around that room at a dozen or so people seized with despair as they listened to this otherwise lovable old man throw in the towel on behalf of humanity.[16] Lovelock's perspective may be extreme in its hopelessness, but you get the picture — we're heading in the wrong direction on energy and need to change how we use it and where we get it.

Harvard University professor John Holdren, science and technology advisor to U.S. President Barack Obama and a former president of the American Association for the Advancement of Science, put it succinctly when he outlined the three choices we have in our faceoff against climate change: mitigation, adaptation, or suffering. "We're going to do some of each," he said. "The question is what the mix is going to be. The more mitigation we do, the less adaptation will be required and the less suffering there will be."[17] The need to reduce our reliance on fossil fuels, become more efficient in how we use energy, and increase our use of low-carbon technologies — the core part of any mitigation strategy — has sparked an era of energy innovation that even Nikola Tesla would find unimaginable. "Scientists, engineers, and entrepreneurs across the globe are responding with unprecedented innovation," according to Christopher Flavin, president of the Worldwatch Institute, an energy and environmental think tank based out of Washington, D.C. "Overnight, the energy business has begun to resemble the IT industry more than it does the energy industry of the past."[18]

Another motivating factor relates to energy security. There's a growing recognition out there that the fossil fuels we have come to depend on to power our economies are going to become more expensive and, from a price perspective, more volatile. I don't think we'll ever run out of fossil fuels, even though they are non-renewable. That's because the cost of finding and extracting and bringing them to market is only going one way: up. We'll simply start using less and less as they lose their competitive edge over alternative energy sources or technologies that help us use energy more efficiently. Creating climate policies that put a meaningful price on carbon will only accelerate this transition. "It should be apparent by now that the future is not going to look like the present. It simply cannot," wrote geoscientist J. David Hughes, who spent 32 years as a scientist with the Geological Survey of Canada.[19] Take oil — the cheap, easy-to-drill stuff is running out, and increasingly we're relying on the more expensive sources that are harder and more energy-intensive to extract, where they're not kept off limits to exploration. With China, India, and other emerging economies jacking up demand for fossil fuels, our current situation is not sustainable. "The party's coming to an end," warned Hughes.

This battle against climate change and concern over the rising volatility of fossil fuel markets, together with projections that the world's population will reach nine billion by 2050, have major economic implications. One, which British economist Nicholas Stern drew attention to in 2006, is the cost of inaction that under a worse-case scenario would amount to trillions of dollars of lost global GDP.[20] The cost of action would be small by comparison, Stern concluded. The second economic implication relates to the new industries and technologies — and jobs — that will be created as we tackle these growing problems. Already, countries are jockeying for position to become global leaders in a new "green economy," and clean technology is the world's fastest growing investment segment. "The green economy is poised to be the mother of all markets, the economic investment opportunity of a lifetime, because it has become so fundamental," Lois Quam, founder of strategic consulting firm Tysvar and former managing director of venture capital firm Piper Jaffray, told *New York Times* columnist Thomas Friedman. "To find an equivalent economic transformation, you have to go back to the Industrial Revolution."[21]

Clearly, there's never been a greater need for new ideas and risk-taking, even in the presence of what may seem impossible or unlikely. Tesla, if he lived today and wasn't trapped in a mental institution, would have been in his element. Does this "need" mean the transition to clean energy sources and technologies will come faster than past energy transitions? It remains to be seen. What is becoming evident is that the energy transition currently in play isn't about moving from one dominant fuel or technology to another; it's about moving from a handful of dominant sources to hundreds. "My strategy on energy technology is to build robustness, to build a portfolio," says SDTC's Whittaker. "Is there a single technology that's going to save us? I wouldn't count on it. It's got to be a whole bunch of little things." We like to think about silver bullets, but, to borrow what is perhaps an overused analogy, we should be thinking about silver buckshot — small projectiles moving together and capable of hitting a much wider target. And we don't need to wait for new breakthroughs before we pull the trigger. Much can be accomplished over the next two decades by more aggressively deploying technologies we have today, including wind, co-generation, geothermal, solar-thermal power, solar photovoltaics, all-electric and hybrid-electric vehicles, second-generation biofuels, and ground-source heat pumps. (For context, discussion of some of these technologies is included in the chapters that follow.) Just as important are the policies needed to support their widespread deployment. "You need intelligent government regulations infinitely more than you need a massive effort to find breakthrough technologies," says climate blogger Joseph Romm, who was acting assistant secretary of the U.S. Department of Energy during the Clinton administration.[22]

I agree with Romm: the search for breakthroughs shouldn't distract us from what we can and must accomplish now. At the same time, the hunt for true breakthroughs, as rare as they may be, is still necessary to sustain us over the long term. We need both leaps *and* incremental steps. The individual efforts profiled in this book may prove an "impossible waste of time." They may lead to dead ends. Whatever the outcome, there is immense value in the journey. The left-behind morsels of innovations won't necessarily go to waste. They can be picked up and used by others who embark on their own ambitious journeys. But there's also a chance these efforts will lead to triumph. And perhaps years, but likely decades, from now we or our children will know, in hindsight, the degree to which these innovators' labors improved our lives.

Interviewed for an article that appeared on August 22, 1937, in the *New York Herald Tribune*, Tesla — 81 years old at the time, less than six years before his death — looked back on his life and seemed quite satisfied that he had repeatedly proved his doubters wrong. "They laughed in 1896 . . . when I told them about cosmic rays. They jeered 35 years ago when I discovered the rotating field principle of alternating currents. They called me crazy when I predicted the radio. And when I sent the first impulse around the world, they said it couldn't be done."[23]

They have often been wrong.

NOTES:

1 MIT press release (April 24, 2001) announcing Raymond Damadian's lifetime achievement award.

2 Lev Grossman, "Forward Thinking," *Time*. October 3, 2004. http://www.time.com/time/covers/1101041011/story.html.

3 Margaret Cheney, *Tesla: Man Out of Time* (New York: Simon & Schuster, 2001), 40.

4 Cheney, 44.

5 Jill Jonnes, *Empires of Light: Edison, Tesla, Westinghouse, and the Race to Electrify the World* (New York: Random House), 203.

6 Tesla also claimed the possibility of communicating with Mars and other planets, as was eventually demonstrated through radio control of the Mars Rover. "That we can send a message to a planet is certain, that we can get an answer is probable," he wrote in his famous essay "The Problem of Increasing Human Energy," published in *Century Illustrated* (June 1900).

7 Speech given by documentarian Robert Uth on June 26, 2006, to the Serbian Unity Congress, in commemoration of Tesla's birthday.

8 Jonnes, 355.

9 Cheney, 258–266; Uth speech.

10 Tyler Hamilton, "Harnessing the energy in oceans and lakes," *Toronto Star*. April 19, 2010. http://www.thestar.com/business/cleanbreak/article/796680.

11 Tim Swartz, *The Lost Journals of Nikola Tesla* (New Brunswick, N.J.: Global Com-

munications, 2000), 17, 115.

12 David Fridley, "Nine Challenges of Alternative Technologies," *The Post Carbon Reader Series*, Energy Issue (August 2010): 1.

13 International Energy Agency, "World Energy Outlook 2009."

14 National Oceanic and Atmospheric Administration, "State of the Climate in 2009." July 28, 2010.

15 U.S. National Research Council, "Advancing the science of climate change." May 19, 2010. (The National Research Council is the operating arm of the National Academy of Sciences and National Academy of Engineering.)

16 Lovelock told the U.K.'s *Guardian* (March 29, 2010) that humans aren't clever enough to deal with climate change and that the failures of democracy partly explain our human inertia on the issue. His comment about cleverness is not a reflection on our ability to invent and innovation. Rather, he's being critical of the institutions we have created, which are ineffective in dealing with an issue as complex as climate change.

17 Oxfam, "Adapting to Climate Change." May 29, 2007.

18 Christopher Flavin, "Renewable Surge Despite Economic Crisis," WorldWatch Institute. May 15, 2009. http://www.worldwatch.org/node/6111.

19 J. David Hughes, "The Energy Issue," *Carbon Shift* (Toronto: Random House Canada, 2009), 58–96.

20 Nicholas Stern, "The Stern Review on the Economics of Climate Change," U.K. government publication. October 30, 2006.

21 Thomas Friedman, *Hot, Flat, and Crowded* (New York: Farrar, Straus and Giroux, 2008), 172.

22 Joseph Romm, "Breaking the Technology Breakthrough Myth . . . ," ClimateProgress .org. April 9, 2008.

23 John J. O'Neill, "In the Realm of Science: Tesla, who predicted radio, now looks forward to sending waves to the Moon," *New York Herald Tribune.* August 22, 1937.

CHAPTER 1

More Bang for the Buck?

A Quicker Path to Nuclear Fusion

*"This is the ultimate greenhouse-gas
reduction doo-dad if we can pull it off."*
— Doug Richardson, CEO of General Fusion

Nuclear fusion is keeping you alive. It's keeping everything alive — plants, bugs, bacteria, viruses, algae, fish, elephants, everything. Just look to the sky on a sunny day and you'll see the source. It's called the sun, which in essence is just an unfathomably massive nuclear fusion machine. Its gravitational pull is so intense that the pressure exerted on its inner core of gases — mostly hydrogen — keeps the temperature at around 15 million degrees Celsius, more than hot enough to trigger reactions that fuse hydrogen nuclei into heavier helium nuclei. When that happens, things go boom. The sun emits the resulting energy in the form of electromagnetic radiation, which travels 150 million kilometers before bathing Earth with life-sustaining sunlight. Without it, we are not. With just the right amount, life thrives.

The idea of somehow recreating this process here on Earth to generate near limitless amounts of safe, emission-free power remains a dream after more than half a century. We understand the theory. We know it can work. We've successfully triggered nuclear fusion reactions on this planet. What we haven't figured out is a way to tightly control it. The nuclear fusion we can achieve today is an instrument of death, not of life. It's called a two-stage thermonuclear bomb, and its design is intentionally uncontrolled.[1] We got our first experience of its destruction in 1952 during a U.S. nuclear test called Operation Ivy; an experimental bomb code-named Mike unleashed 10.4 megatons of explosive energy on a small island in the middle of the Pacific Ocean. That island, Elugelab, was entirely obliterated. The only evidence of its past existence is a two-kilometer-wide underwater crater. Scary.

But what if we could tame the fusion beast for the good of the globe? What if we could harness its power to create clean energy that is affordable, safe to produce, and which doesn't solve one problem by creating others, such as the proliferation of nuclear weapons or a legacy of highly radioactive waste that remains dangerous for hundreds of years? And if we could, what would the effort look like? A high-profile international project with hundreds of lab-coat wearing scientists from several countries collaborating under a single umbrella? Billions of dollars, of course, would be required for such a monumental global effort and, if all went well, fusion power would begin feeding the electrical grid in about three, or more likely four, decades. If all this sounds familiar, it's because it's a general description of the controversial International Thermonuclear Experimental Reactor project, better known by the acronym ITER.

ITER has been a going concern for more than 25 years. It began in 1985 as a research effort shared between the United States, Japan, the former Soviet Union, and what is now the European Union, but a detailed engineering design that was acceptable to all parties didn't emerge until 2001. It wasn't until the fall of 2006 that members of ITER — now consisting of the EU, China, Japan, Russia, the United States, India, and South Korea — struck a formal agreement to move forward with the 10-year construction of an experimental facility in the south of France that could demonstrate just the *feasibility* of fusion power. The original cost of construction was pegged at 5 billion Euros, or $6.4 billion U.S. By fall 2010, it had ballooned to more than 15 billion Euros, or nearly $20 billion U.S.[2] This enormous cost is what's earned ITER the label "controversial." It has been hobbled by United Nations–style bureaucracy and, without anything yet built, has seen its cost estimates triple in just four years. Some observers of the project, from both academia and industry, rightly wonder whether this is a worthwhile international investment or just a makeshift project for career nuclear scientists. Could $20 billion be better spent elsewhere? Can we afford to wait decades for the panacea of fusion power as we stand on the precipice of climate catastrophe? "I think [fusion power] will be solved," British scientist and climate expert James Lovelock once told me. "I just don't think it will be in time."

Michel Laberge shakes his head like a disapproving mother when he talks about ITER. "ITER is a dinosaur," he said. "It is so complex with all those countries involved. It's not moving at all." Laberge is founder and president of General Fusion, a small company located in a nondescript industrial mall less than an hour outside of Vancouver, British Columbia. You might call General Fusion the anti-ITER. This gutsy little company, armed with three dozen or so engineers, aims to demonstrate by 2014 that it can create a "hot" nuclear fusion reaction that gives off more energy — much more energy — than it takes to trigger it. It plans to do this with $50 million, not $20 billion. And it expects to be supplying fusion-based power to the grid by 2020, not 2040 or later. Bold, ambitious, potentially disruptive, and, as you'll find out, based on sound science. But try knocking on the doors of investors to raise money for such an endeavor, or ask someone in the tightly knit community of government fusion scientists for a helping hand. "They tell us it's totally impossible and that we're completely crazy. That's the consistent message," Laberge said. "From the scientific world, the economic world, the government — all of them — they just don't take us seriously. Their argument is as follows: we have billions of dollars, we have thousands of the brightest physicists working on this thing for years and years, and you and your little bunch in the boonies with 50 million dollars are going to make it? That's totally impossible, you flakes. Go away. We don't want to hear about it."

DODGY TRACK RECORD

They don't want to hear about it because a number of high-profile fumbles have left a big embarrassing blotch on the history of fusion science. In 1958, a Nobel prize–winning scientist from the United Kingdom named Sir John Cockcroft gathered the nation's media to announce that he and his research team had demonstrated controlled nuclear fusion in a giant machine nicknamed Zeta. Cockcroft beamed with pride. "To Britain," he declared, "this discovery is greater than the Russian Sputnik."[3] Commercially generated fusion power was two decades away, he predicted. Just months later Cockcroft was forced to admit that his observation of nuclear fusion and subsequent claims were an unfortunate mistake. Fast-forward 31 years to the infamous "cold fusion" claims of chemists Stanley Pons and Martin Fleischmann. Pons, chairman of the chemistry department at the University of Utah, and Fleischmann, a veteran professor of electrochemistry at the U.K.'s University of Southampton, had collaborated for several years when, in 1983, their Utah lab experiments demonstrated a nuclear fusion reaction at room temperature. It was a relatively simple setup: a cathode, in this case a rod made of the precious metal palladium, was inserted into a glass tube filled with heavy water, which is water with a high amount of a hydrogen isotope called deuterium. When electricity was applied to the cathode, it caused bursts of heat that Pons and Fleischmann believed to be a fusion reaction — the creation of new atoms caused by the fusing of deuterium in palladium. The output of energy, according to the two professors, was substantially higher than the energy going into the process in the form of electricity. They continued

to refine their experiments until 1989 when university admin-
istrators, bursting with excitement about this ground-breaking
discovery, jumped the gun and scheduled a March 23 news
conference to announce this world-changing breakthrough.

Standing behind a podium in a university auditorium packed
with journalists — a scene captured on video that, thanks to the
wonders of the Internet, is now available on YouTube — Pons
seemed quite confident about the discovery. "The heat we can
measure can only be accounted for by nuclear reactions," he said.
"The heat is so intense it cannot be explained by any chemical
process that is known." He went on to paint a very positive portrait
of the future. "I would think it would be reasonable within a short
number of years to build a fully operational device that could
produce electric power or drive a steam turbine." Fleischmann,
acknowledging that much more research was needed to establish
a scientific base for their findings, was equally optimistic about
the "possibility of realizing sustained fusion with a relatively
inexpensive device." A global fusion frenzy immediately ensued,
with laboratories around the world attempting to replicate the
Pons and Fleischmann setup. The state of Utah even coughed up
$4.5 million to establish a National Cold Fusion Institute in Salt
Lake City.[4]

But something didn't smell right, and by the end of 1989 it
had become clear that the claims were nonsense. All attempts at
replication had failed or had been proven faulty. In November
1989, an advisory panel created by the U.S. Department of Energy
reported that the Pons and Fleischmann experiments and
attempts to replicate them "do not present convincing evidence
to associate the reported anomalous heat with a nuclear process."
Nuclear fusion at room temperature, the panel wrote, "would be
contrary to all understanding gained of nuclear reactions in the
last half century; it would require the invention of an entirely
new nuclear process."[5] The whole affair was a near death blow to
the reputations of both men, who were accused of acting unethi-
cally. At best, their work was shoddy. They were lampooned. They
ended up moving (fleeing?) to France in 1992 to continue their

work at a privately funded lab. Pons never returned to the United States and ended up becoming a French citizen. Fleischmann left France in 1995 and returned to England where, for the next decade, he dabbled in cold fusion research.

As far as dabbling goes, there is actually a fair amount of cold fusion experimentation going on in the scientific community, despite the fact that the U.S. patent office has stopped granting patents to cover such work.[6] One champion of cold fusion is Peter Hagelstein, an associate professor of electrical engineering at the Massachusetts Institute of Technology. He continues to carry the cold-fusion baton as part of a shrinking group of dedicated researchers who are treated like pariahs by mainstream academics and scientists. One of the most recent demonstrations of cold fusion was conducted by Yoshiaki Arata, a professor of physics at Osaka University in Japan, in spring 2008.[7] It had blogs buzzing but the mainstream media stayed far away.

The media, with the exception of the Italian press, has also steered clear of scientist Andrea Rossi and his partner Sergio Focardi, a physicist and professor emeritus at the University of Bologna. In January 2011, the two men demonstrated their own cold fusion apparatus, which they claim fuses nuclei of nickel and hydrogen to produce copper and huge amounts of excess energy. Rossi is apparently building a one-megawatt plant at his own expense for a company in Greece. A demonstration of that plant is expected in late 2011.[8] But beyond fringe observers, will the rest of the world be watching?

The perception that low-cost fusion is a fool's game — a perception General Fusion must contend with every day — was further strengthened by a scandal related to another tabletop process, this one called "bubble fusion." Rusi Taleyarkhan, a professor of nuclear engineering at Indiana's Purdue University and former scientist at the U.S. government's Oak Ridge National Laboratory, published a paper in 2002 describing how he had aimed high-frequency sound waves at a glass flask filled with a deuterium-rich liquid. Pressure created from these ultrasonic waves, which can't be detected by the human ear, caused tiny bubbles in the liquid to violently collapse and release a tremendous amount of heat. Researchers of bubble fusion refer to the setup as "star in a jar."[9] Following this initial experiment, there were several research groups seriously exploring bubble fusion, but by 2006, and after burning through millions of dollars in funding, none of them were able to replicate Taleyarkhan's demonstration. It was around that time when *Nature*, the highly respected scientific journal, published an investigation of Taleyarkhan's work and concluded that "the circumstances surrounding the experiments reveal serious questions about their validity."[10] The journal's investigation created such a stir that Purdue University launched its own probe and, in 2008, ended up finding Taleyarkhan guilty on two counts of alleged research misconduct; his research chair was subsequently taken away.[11] So what was his misconduct? Taleyarkhan led people to believe that the bubble fusion effect he observed had been independently verified when, in fact, the "verification" had been conducted in Taleyarkhan's own lab. The U.S. Office of Naval Research, which funded part of Taleyarkhan's work, called the professor's actions "research fraud" and barred him from receiving any sort of federal funding until 2012.[12]

To be clear, there is serious and important work going on in the areas of cold fusion and bubble fusion. Unfortunately, the blunders and bad apples of yesteryear have pushed what was already considered fringe science into an even more defensive posture, and researchers in the field are akin to lepers at a beauty pageant.[13] General Fusion is thrown into the same category despite the fact that its approach doesn't involve tabletop setups, flasks, or collapsing bubbles. The foundation of its reactor technology was first developed in the mid-1970s by the U.S. Naval Weapons Research Lab and is well supported by known scientific theory. But because General Fusion is a small operation, because it's not an international effort backed by billions in government dollars, and because memories of Pons and Fleischmann and Taleyarkhan remain so vivid in the minds of so many, it has been branded another fusion fly-by-night that's not worthy of attention or investment. "Nobody takes us seriously," said Laberge.

OLD BOYS' CLUB

Should General Fusion be taken seriously? Let's take a moment to examine what the company is trying to do, and how that differs from the two generally accepted approaches to nuclear fusion today: ITER's magnetic fusion approach and the inertial confinement fusion work being done at the U.S. government's National Ignition Facility in California. At the heart of ITER's project is an imposing machine called a tokamak, which, when completed, will weigh in at 23,000 tonnes — half the weight of the *Titanic* and roughly equivalent to 67 fully loaded Boeing 747 jumbo jets. A tokamak is essentially just an electromagnet shaped on the inside like a hollowed-out doughnut. It's made of 48 magnet systems that, together, will create a magnetic field that's 200,000 times greater than what the Earth produces. The idea is to inject a gas-like plasma of deuterium and tritium into the cavity of the electromagnetic doughnut. Intense magnetic fields emanating from the walls of the cavity will keep the electrically charged plasma under control and at enough distance from the walls to prevent damage to equipment.[14] The plasma will have nowhere to go, but because it will be energized it will race around inside the cavity of the doughnut like a dog chasing its tail. Before fusion can occur, however, the plasma must be heated up substantially — and by substantially I mean 10 times hotter than the core of our sun. This will be done by subjecting the plasma to a barrage of radio waves, electromagnetic radiation, and "neutral" beams.[15] "The plasma will catch fire when it gets hot enough. It's called ignition," said Laberge, explaining the point at which the deuterium and tritium nuclei in the plasma begin fusing to form helium nuclei. At that stage, the trick is to simultaneously continue magnetically containing the plasma, harvest the heat for power generation, *and* keep the reaction sustained.

If it works as envisioned, the scientists at ITER hope that the 50 megawatts of power being used to drive and control the reactions will results in 500 megawatts of power coming out — a 10-fold gain in energy. It's an ambitious target, considering we've yet to even reach a break-even point in fusion research. The best effort to date has been at the ITER predecessor JET, the Joint European Torus facility, which in 1997 achieved an output of 16 megawatts from an input of 24 megawatts. There are a number of other small tokamak experiments being conducted around the world. In May 2010, for example, Russia and Italy agreed to push forward on their own project, Ignitor; China is also working on its own machine. But ITER, with its size and $20-billion price tag, is the top dog. Laberge described it as scientific overkill that is being bogged down by international politics and bickering over the bill: "We have a joke here that China will produce a tokamak the size and power of ITER before ITER will have the toilet installed in its building."

The second main approach being studied in fusion physics is called inertial confinement, with most of the work being conducted at the U.S. government's Lawrence Livermore National Laboratory (LLNL). One of nine national labs spread throughout the country, LLNL houses the largest laser facility in the world, the National Ignition Facility (NIF), which came to life in March 2009. It's a controversial operation, mainly because its mandate is to assist in the stewardship of nuclear weapons through controlled testing that can offer insight into the physics behind thermonuclear explosions. The ITER-like hunt for commercial fusion power, critics say, is a secondary pursuit that is used to put a peaceful spin on weapons research. No wonder its $3.5-billion price tag and ongoing costs are covered under the U.S. government's nuclear weapons budget. Still, the facility's laser system is an impressive feat of engineering, consisting of 192 powerful laser beams carefully positioned within a building the size of a football stadium. The whole operation is an orchestra of technology that demands perfect execution.

To achieve ignition, the plan is to aim the laser beams at a spherical fuel pellet about one centimeter in diameter. The pellet has a glass shell and inside it a densely packed mixture of deuterium and tritium. When the lasers strike, they do so in one billionth of a second, creating an instant burst that compresses the pellet and produces the heat and pressure that triggers a fusion reaction. Voilà! You've got ignition. As with ITER, scientists hope they can show an output of energy at least 10 times more than the energy required to power the lasers, though hundreds-fold gains are likely necessary to demonstrate that commercial fusion power production using this expensive method is economically viable.

LLNL does a good job of explaining why 192 ultraviolet lasers are needed with the analogy of a water balloon being squashed between two hands. "No matter how hard you try to spread your fingers evenly over the surface of the balloon, it will squirt out between your fingers. Many more fingers would be needed to compress the balloon symmetrically," it explains on its website. "NIF's designers arrived at 192 focused spots as the optimal number to achieve the conditions that will ignite a target's hydrogen fuel and start fusion burn." Those lasers must be positioned and triggered with near-absolute precision. First, they must hit the target at the same time. If one is more than 30 picoseconds, or 30 trillionths of a second, out of synch the entire experiment fails. They must also strike their targets within 50 micrometers, or half the width of a human hair. The fuel pellet itself is slightly larger than a pearl on a necklace. The laser system's pointing accuracy, according to LLNL, "can be compared to standing on the pitcher's mound at AT&T Park in San Francisco and throwing a strike at Dodger Stadium in Los Angeles, some 350 miles away." And the system packs a powerful punch. For the few billionths of a second it is pulsed, its peak power is about 500 times the power used by the entire United States at any single point in time.[16] It's all very interesting, and tests conducted in 2010 are progressing well. Scientists at the National Ignition Facility may indeed achieve ignition over the coming years, and in doing so bring a much-needed boost of confidence to the area of fusion research. Unfortunately, the approach may prove too expensive, inefficient, and complex to be practical for commercial fusion power generation.

HYBRID VIGOR

General Fusion, unlike ITER or NIF, is pursuing an approach called magnetized target fusion, which it considers a best-of-both hybrid of magnetic fusion and inertial confinement fusion. ITER's approach, for example, confines the plasma but uses externally supplied heat instead of compression to achieve thermonuclear conditions; the National Ignition Facility uses a sudden burst of lasers to compress the plasma, but can't confine it beyond a billionth of a second. Magnetized target fusion, in theory, achieves a nice balance of both compression and confinement, and in doing so makes it possible for General Fusion to strip a lot of cost, complexity, and inefficiency out of the demonstration reactor it's hoping to build. No super-expensive lasers. No enormous superconducting magnets that cost billions of dollars. Perhaps what's most unbelievable is the company plans to demonstrate this ideal balance of compression and confinement on a budget that might just cover the cost of building and maintaining ITER's washroom facilities over the next 20 years.

Here's how it will work: imagine a big metal globe, a few meters across, that's filled with a mixture of liquid lithium and lead. The liquid is injected into the globe in a way that causes the mixture to spin. This creates a vortex, similar to the whirlpool that forms when one rapidly stirs a glass of water with a spoon. The vortex in the globe stretches from top to bottom, creating a pathway — an air gap — right through the middle of the mixture. Two separate large metal devices, called plasma injectors, are connected to the globe. Shaped like the top of a sharpened pencil, one plasma injector is attached by its pointy end to the top and

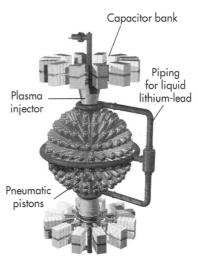

Capacitor bank

Piping for liquid lithium-lead

Plasma injector

Pneumatic pistons

one to the bottom of the globe. There's a good reason for the shape. The plasma injector's job is to take tritium-deuterium gas, turn it into a doughnut-shaped plasma called a spheromak, and then push the plasma doughnut through the increasingly narrow part of the injector, a process which compresses the plasma into a smaller and smaller doughnut until it reaches the tip of the injector. "It compresses the plasma, which heats it and makes it more dense as it goes through," explained General Fusion chief executive Doug Richardson. The process, which is happening in two injectors at the same time, relies on massive banks of capacitors and batteries to create the plasma and the magnetic fields that shape it, as well as on the electromagnetic forces required to push and compress the spheromak as it moves through the injector. When the spheromaks are at the tips of the injectors, the fusion fireworks are ready to begin. One spheromak is launched into the top of the metal globe and one is launched into the bottom. They travel toward each other through the air gap in the spinning lithium-lead liquid until they meet each other at the center.

On the exterior surface of the metal globe, there will be 200 to 300 pneumatically controlled pistons jutting out. If you remember the character Pinhead from the 1987 Clive Barker horror movie *Hellraiser*, you have a very rough idea of what this contraption will look like. When the two spheromaks meet in the middle of the metal globe, the pistons are simultaneously triggered. At a velocity of 100 meters a second, they ram the surface of the metal globe, creating sound waves that accelerate through the lead-lithium mixture and converge toward the center. The sound waves gather so much speed that they become more powerful shock waves before impacting the plasma. When they do impact, the spheromaks rapidly compress and heat up to conditions that trigger a nuclear fusion reaction. Now comes the part where General Fusion makes electricity out of all this. As we know from discussing ITER, the fusion of deuterium and tritium into helium releases a starburst of energy. In the General Fusion scenario, that energy is absorbed into the surrounding lithium-lead liquid. That liquid is pumped out of the globe, and the heat within it is extracted and used to make steam for power generation.

Richardson told me a big risk in this process is that nobody has ever compressed a spheromak to the point where a fusion reaction is triggered. "There's no reason why it won't work," he said. "But nobody has ever proven it." Not only does General Fusion have to prove it can be done, it has to eventually design a commercial machine that replicates the process every second — that is, every second two spheromaks must be shot with precision into the heart of the globe and compressed to the point of achieving fusion. That's 60 shots every minute, or nearly 32 million successful fusion reactions every year. Throughout this, the superheated lithium-lead mixture must be constantly pumped out, extracted of heat, and then pumped back in. "The way to think of this overall is as a thermonuclear diesel engine," Richardson explained. "You inject fuel. Then you squish it, which is exactly what a diesel engine does. And if it gets hot enough, it burns."

So unlike ITER, a project designed to confine and sustain a single nuclear fusion reaction for years — even decades — General Fusion's approach is to create smaller and easier-to-control fusion pulses every second. And unlike the National Ignition Facility, which uses laser pulses, General Fusion's approach is mechanical. "They're literally using a sledgehammer," said Rick Whittaker, vice president of investments at granting agency Sustainable Development Technology Canada, which has stuck out its head to become the company's biggest single funder.[17] General Fusion believes it can build a commercial fusion power plant that delivers 40 units of energy for every single unit that goes into it, four times higher than what ITER is aiming to achieve. A fraction of the energy coming out would be harvested to trigger the next reaction, keeping the process self-sustaining. "What we want to have at the end are economic power plants, not just a device that proves net gain for fusion," said Richardson. Those plants would likely be rated about 300 megawatts, or the equivalent of a small coal-fired power plant, and the fusion reactor itself would be about the size of a large barge. The company claims its fusion plants would eventually be capable of producing electricity for less than 5 cents per kilowatt-hour, making it one of the cheapest and cleanest forms of power generation on the planet.

If magnetized target fusion is such a good idea, then why aren't the top fusion scientists in the world working on it? Why aren't investors knocking down General Fusion's door? Why are Laberge and his team written off as crackpots?

A BRILLIANT MIND

Hoping to answer such questions, I traveled to Vancouver in the summer of 2010 to get a closer look at General Fusion and a better appreciation of its struggle. When I first sat down with Michel Georges Laberge and told him I was dedicating a chapter of this book to his work on nuclear fusion, the Quebec-born engineer, wearing a blue T-shirt, green khakis, and purple-tinted sunglasses, asked me what I planned to call the book.

"*Mad Like Tesla*," I said, explaining that inventors and entrepreneurs doing disruptive work in the energy field, like himself, have much in common with inventor Nikola Tesla, despite their century of separation.

"Oh shit," replied Laberge in a French accent that remains thick

Doug Richardson (left) and Michel Laberge pose in front of their prototype plasma injector, surrounded by capacitor banks.

even after 25 years outside his home province. "Tesla finished his life in debt and destitution. Is this my future?"

I read a clever description of Laberge in a *Popular Science* article about General Fusion that appeared in early 2009. "On the mad-scientist appearance scale, Laberge is maybe a 4 out of 10; he's a little rumpled and wears out-of-style wire-rimmed eyeglasses," wrote journalist Josh Dean. "But get him a little agitated and he starts to tug at his hair and slips to maybe a 5 or 6."[18] As long as Laberge doesn't start seeking the love of a white pigeon, I thought to myself, he'll be okay.

Laberge would have been far better off financially had he stayed on the path he was on. During the 1990s he was a senior physicist and principal engineer at Vancouver-based Creo Products, a developer of digital laser technologies for the print-ing industry that ended up being purchased by Eastman Kodak in 2005. His skills were appreciated, his job was stable, and the money was quite good. Laberge's expertise was in making sure the laser beams on commercial printers were properly aligned; if they're not, strange stripes called microbands appear on the printed stock. Not the most exciting job, but it paid the bills.

In 2000 Laberge's role at Creo took an unexpected turn. Amos Michelson, the company's chief executive at the time, was in awe at how much money was changing hands during a telecommunications boom that had companies such as Nortel Networks and Cisco Systems on irrational multibillion-dollar shopping sprees. One deal saw Nortel Networks spend $3.4 billion in stock for a tiny Silicon Valley company called Xros, which had zero revenue and only a few dozen employees. Xros had developed a "micromirror router," a type of optical switch that used thousands of tiny mirrors to reflect light pulses sent through fiber-optic communications lines. It allowed the big phone and Internet companies to dramatically boost the capacity of their networks. "My CEO says, 'Wow, $4 billion [Canadian] for 30 guys in a garage, what a fantastic evaluation! Let's do an optical switch of our own and throw our share price through the roof,'" recalled Laberge. "So I thought, okay, I think I can do this."

It was an odd project for a printing company, but Laberge was assigned to lead a crack team of five engineers who reported to a project manager named Doug Richardson (who was later recruited to be General Fusion's CEO), and within six months and for less than $1 million they produced an optical switch that outperformed anything else on the market. "This inflated my ego a little bit," said Laberge, who took great satisfaction in knowing that for dramatically less money and time he was able to beat engineering teams at Nortel, Cisco, and AT&T. Unfortunately, just as Creo was preparing to introduce its superior optical switch to the marketplace and stir up a bidding war for its technology, the telecommunications market collapsed. Investment plunged. Orders evaporated. Companies such as Nortel Networks eventually became worthless penny stocks. "Nobody was interested in telecom whatsoever, so my big project went into the garbage. It was terminated," said Laberge, scratching a red beard that's sprouting much more grey these days. "They ended up putting me back into laser printers."

Having tasted something different, something more exciting

through the fast-paced telecom project, Laberge was reluctant to reclaim his job as the microband killer. "I was bored of this, and faced with going back I said to hell with it. I was turning 40 and had a bit of a mid-life crisis. Most guys get a Porsche, but I decided to throw my job away and form a nuclear fusion company." He took the summer off and spent time on the deck of his house on heavily forested Bowen Island, a nature lovers' paradise just a couple of kilometers west of the B.C. mainland. It was there, with mountains in the background, ocean in all directions, and an Internet-connected laptop at his side, that he narrowed down the approach to nuclear fusion that would result in the birth of General Fusion. The company's first "headquarters" was an old, ramshackle gas station garage near his home.

It's not often that within a period of two years a person moves from laser printers to telecom switches to nuclear fusion, but Laberge is among a select group of individuals with the qualifications to do it. During the early 1980s he studied physics at Quebec's Laval University, but left his home province in 1985 to pursue a Ph.D. in fusion physics at the University of British Columbia. He earned that degree in 1990 and, just before joining Creo, continued post-doctorate work at the École Polytechnique in Palaiseau, near Paris, and at Canada's National Research Council in Ottawa. His work at Creo was a complete departure from what he'd been studying, but it also made him an expert in the digital signal processing (DSP) and servo-control technologies that were central to the development of high-speed laser printers, and which would prove a crucial element of General Fusion's reactor technology. "Michel Laberge is one of the few kinds in the world who understood the engineering behind digital signal processing and plasma physics," said one of General Fusion's earliest investors, venture capitalist Michael Brown, during one of our many chats together. "Hardly anybody has that combination."

It's a vital combination, one that explains why Laberge could approach fusion with an open mind and from a fresh perspective. Most fusion scientists have spent their entire careers pointed in the same direction and are not prepared to shift course. They have tunnel vision, and this, as far as Laberge is concerned, has created an environment of science-by-committee where advances in fusion have been incremental, at best. "All the funds and grants are distributed by committees that are full of physicists who work on the subjects being funded," Laberge told me. ITER has been a going concern for nearly three decades. Once the decision was made to head in the direction of magnetic fusion, the project became a slow-moving train that over time has gathered huge momentum. About 1,000 people will be employed directly by ITER during its operational phase. Fusion research is a small niche within the larger nuclear research field, yet ITER allows hundreds of these specialized researchers to build careers out of this single project. The same can be said of the National Ignition Facility. Abraham Loeb, director of the Institute for Theory and Computation at Harvard University, writes that this form of groupthink is becoming more prevalent in basic scientific research. Scientists, he believes, are "increasingly pursuing projects in large groups with rigid research agendas and tight schedules that promote predictable goals."[19] Unlike most fusion scientists, Laberge didn't carry any such baggage. He wasn't on the ITER train.

That rare perspective allowed him to rediscover magnetized target fusion, which the U.S. Naval Weapons Research Lab had worked on back in the 1970s. They came up with a design, called LINUS, which is the underlying basis of General Fusion's reactor design. LINUS was shelved, not because it couldn't work, but because the technology didn't exist at the time to build one that could work. Recall that this is a mechanical process, relying on hundreds of huge pistons striking a metal ball at the exact same time in a microsecond. The technology didn't exist in the 1970s to control the timing of those piston strikes with the kind of precision and speed that could achieve a fusion reaction. Like many inventions, it was abandoned because pieces of the puzzle weren't just missing, they didn't exist at all. The problem is that when those pieces came into existence, they remained missing, because anyone who once cared had moved on, or even passed away.

Nobody was looking to breathe new life into magnetized target fusion, until Laberge came along in 2001. In the case of LINUS, the missing pieces were digital signal processing (i.e., superfast computers) and high-precision servo-control systems, the same technologies that Laberge built a career on while at Creo. Laberge realized that by employing technology he knew intimately, LINUS now had a good chance at success. History is littered with inventions that were rediscovered later when advances in material science, digital processing, and other technologies made them possible; Leonardo da Vinci's helicopter is just one popular example. Patent offices are a treasure trove of unrealized potential if one stops to scour their archives.

ENTER AN ANGEL

But let's get back to venture capitalist Michael Brown and his impressions of Laberge. In many ways Brown is the father of Canadian venture capital, particularly when it comes to energy technologies. He's been at it since 1968, when he co-founded Vancouver-based Ventures West Capital. In 1987 he was the first venture capitalist to make a bet on Ballard Power Systems, which was the brightest light in the hydrogen fuel-cell market until dreams of a hydrogen economy began fading in the late 1990s. Over the years his money and influence have helped dozens of companies, many of them based in British Columbia, including Canadian aerospace firm MacDonald, Dettwiler and Associates and wireless pioneer Mobile Data International, which was purchased by telecom giant Motorola in 1988. Brown left Ventures West in 1999 and two years later co-founded Chrysalix Energy, a firm that manages more than $300 million in venture investments. Now in his early 70s and perhaps a more contemplative soul as a result, Brown has become somewhat of a philosopher of venture investing.

It was around 2002 when Brown first became aware of Laberge and his seemingly insane mission to build a fusion reactor. He heard of the fusion work through an engineering superstar by the name of Dan Gelbart, who in an earlier career at MacDonald

Dettwiler and as a founder of Mobile Data had become friends with Brown. Gelbart was also co-founder of Creo and knew Laberge quite well. "Gelbart thinks Michel Laberge is one of the smartest engineers he's ever worked with," said Brown. "He spoke highly of Michel, so I decided to throw in $10,000. Then I just sort of stood on the sidelines for a while. I mentally wrote off my investment." It was Brown's personal money; it didn't come from Chrysalix. Laberge also had strong support from friends and family during these early days. He managed to raise about $400,000 from a group of about 40 people, most of them former Creo colleagues. Richardson was one of them, long before he became General Fusion's CEO in 2006.

Laberge's local support network expanded as he reached out to the B.C. technology community. He convinced Gelbart to join General Fusion's board, as well as Denis Connor, who as former head of the Science Council of British Columbia was heavily involved in a number of clean technology ventures. Venture capitalist Jim Fletcher, one of Brown's partners at Chrysalix (and a former partner at Ventures West), also joined the board after investing some of his personal money in the company. Connor and Gelbart are named on one of General Fusion's first patents along with Laberge and Simon Fraser University chemistry professor Ross Hill — the result of a private brainstorming session that Laberge had organized with, in his words, "lots of local high-tech dudes." Connor, Gelbart, and Hill contributed ideas that ended up making their way into General Fusion's initial reactor design.

For a few years Laberge worked away on the concept as a one-man show. It was spring 2007 when Connor, still on the board, bumped into Brown and mentioned that Laberge, who by then had Richardson on as CEO, was making solid progress and coming up with interesting results. Brown was intrigued and became more engaged. After doing extensive due diligence of his own, he decided to take General Fusion under his wing. He knew that if the company was to get anywhere, it needed to raise tens of millions of dollars to demonstrate proof-of-concept and eventually build a device that could achieve net gain — i.e., a fusion reaction that releases more energy than it took to trigger it. Some initial seed money did flow, through investments from Chrysalix, the Business Development Bank of Canada, and another Vancouver-based venture firm called GrowthWorks.[20] "We used to meet in the coffee shop in the Holiday Inn right by the Second Narrows Bridge," a railway bridge that connects Vancouver to North Vancouver, recalled Brown. "We took trips to the venture-capital community and had lots of highs and lows, mostly lows."

Brown spent a lot of time helping the company look more appetizing for investors. He prepared many of their financial presentations, put their budgets into a recognizable format, wrote up a marketing plan, and even drafted a prospective hiring plan for senior management. He also realized the company needed some strong endorsements from engineers and scientists who knew what they were talking about. Brown made use of some contacts he had at aerospace giant Boeing, which was once a limited partner in Chrysalix, and managed to get a team of five Boeing engineers to write up a 14-page technical review that assessed the mechanical control and manufacturing issues of General Fusion's reactor design. Brown was encouraged by their final report: "They basically concluded there's no reason why it isn't going to work." Another analysis was done by leading fusion expert Ken Fowler, professor emeritus of nuclear engineering and plasma physics at the University of California, Berkeley. Fowler has close connections with the Lawrence Livermore National Laboratory and is one of the rare fusion scientists with an expertise in spheromak plasma confinement. When I spoke with Fowler in 2009, he was cautious, but still held high hopes for General Fusion and said there were no technical showstoppers to its approach. "I'm rooting for them," he said.

Armed with these positive reports from two credible sources, and given the game-changing implications of the technology, you would think a seasoned venture capitalist such as Brown would have an easy time tapping into his network and bringing more investors on board. That's not how the story unfolded. "They looked us in the eye and said it can't possibly work. So we said, alright, come meet the guys and talk to the scientists. But they just wouldn't do it; they won't talk to the scientists," said Brown, expressing frustration during one phone call we had in late 2009. "What do we have to prove for people to take it out of the category of junk?" Laberge told me the investment community, by and large, just isn't prepared to understand complex technologies. General Fusion had a detailed 400-page report that anyone could read, but few were willing to take the plunge. "In order to understand this, you need at least a degree in physics, and then at least two to three weeks of reading and researching on websites," said Laberge. "Who is going to take this sort of time out of their job to understand something new? Who is going to throw [away] two weeks of their time to try to understand an offer? Is it easier to say no, go away, you're a flake, or to take two weeks and look into it? It's easier to say go away. So that's the biggest problem we have." Brown said it's partly about Silicon Valley groupthink. "In Silicon Valley, unless the high-profile guys have anointed something, it's like everyone else is afraid to go to a cocktail party and tell someone about a technology like this," he said. "What groupthink gets them to do is to fund 174 solar thin-film companies."

A big boost came in March 2009 when a Canadian agency called Sustainable Development Technology Canada (SDTC) announced that it would grant General Fusion $13.9 million Canadian. The agency was created by the federal government to help fund emerging clean technology ventures, but its commitment to General Fusion was a huge gamble for an organization where the average deal size is closer to $3 million and whose funding comes from a government with no formal fusion research program. Canada, it should be noted, has a stubborn obsession with its homegrown brand of nuclear fission reactor technology: the Canada Deuterium Uranium reactor, known as CANDU. Rick Whittaker, SDTC's vice president of investments, said the agency's support for General Fusion was far from guaranteed: "Because of all the skepticism, we dug deep and we brought in experts." The

SDTC funding ended up diluting investment risk, making it easier for private investors to loosen up their wallets. It was still like pulling teeth, but by August 2009 General Fusion was able to raise $13.75 million more from GrowthWorks and Chrysalix, as well as newcomers Braemar Energy Ventures and the Entrepreneurs Fund. It was enough money to design and demonstrate crucial components, such as the pistons and plasma injectors. But about $25 million more will be needed to build the metal sphere that lies at the heart of the reactor. Those pieces will go together to demonstrate net-gain fusion by 2014. "After that, we'll need multiple billions of dollars to make it into a power plant," said Richardson. "But if we can show that net-gain, the money should be available to go to that next level."

Laberge gives Mike Brown full credit for getting them this far. "Without Mike we don't exist. He introduced us to all the right people." Perseverance is paying off. Investors willing to take the time to study Laberge's approach are slowly coming onside. General Fusion announced in May 2011 it had raised another $19.5 million, with a portion of that — to the surprise of many — coming from Jeff Bezos, the multibillionaire founder of Amazon.com.[21] Canadian oil company Cenovus Energy also contributed, bringing total funding in General Fusion to about $46 million. Sometimes crazy sells, particularly when the risks are far outweighed by the potential rewards.

PLASMA GO BOOM

It was a perfectly sunny July day in Vancouver when I visited General Fusion's office to get a close look inside. The first thing that grabbed me was the color of the office walls, a disgusting shade of green that told me the place employs a bunch of engineers who couldn't care less about aesthetics. Richardson greeted me at the front reception wearing faded blue jeans and a grey T-shirt with a smiling orange sun on the front and the words "A day without fusion is like a day without sunshine." Richardson fits the classic Pacific Coast image, at least from an easterner's perspective. He's in shape, laid back, and looks like he could be a surfer — definitely an outdoorsy type who likes to take his canoe and go portaging. He took me for a tour of the warehouse in the back where all the tinkering happens. The main attraction was the plasma injector, which looked like something out of the *Matrix*. Wires sprouted from it and cords cluttered the warehouse floor. Surrounding it were banks of batteries and capacitors, the latter controlled by Russian-designed switches, which create and drive the spheromaks that are injected into the reactor. Shortly after the first round of funding closed in August 2009, company engineers began building the injector. It took nine months to assemble. "If this injector were to be built in a government lab, it would take four years," Richardson said proudly.

A couple of minutes later our ears were pierced by a loud siren, like something you hear on *Star Trek* whenever a Klingon Bird-of-Prey is about to attack the U.S.S. *Enterprise*. Richardson shouted to a bunch of engineers standing nearby. "Is there a reason we shouldn't be on the floor?" he asked. Without a firm answer, he grabbed me a helmet and some headphones and told me to put them on — just in case. "If it discharges the wrong way, the thing can come apart," he said with voice raised. "A capacitor could fall apart. If you put that much energy into a single piece and it blew, it could send a little bit of shrapnel flying." The siren went silent moments later and the engineers, clearly perplexed, began to troubleshoot. The issue, Richardson explained as we stood behind a military-grade blast wall, had to do with a funny sound heard during the discharge of power to the injector. "If things sound unusual in the circuit, there could be something that's gone amiss. Life has taught us if you hear such things it's better to find out what it is than ignore it." It's a wise approach, when one considers that the capacitor banks when fully powered will carry a 10-microsecond jolt that, if misdirected, could cause equipment to vaporize. On this day, it's using about 40 percent of full power, rising every week.

Richardson walked me to another part of the industrial complex that is used for storage. We saw one of the original pistons designed by Laberge, as well as a "Version 2" to be used in more advanced testing. Beside it was a station where the lead-lithium is heated and kept in liquid metal form. "Before we make the sphere, we have to understand how to handle the liquid metal," he explained. "Do we have all the right seals? Do our seals leak? How many shots can they take? Can they operate at a temperature gradient?" Computer simulations don't cut it, he said. "Investors want it proved. The goal is to de-risk all the major components [using] the least amount of money possible."

Have there been any hiccups, what I refer to as "Oh shit!" moments? "There are little 'Oh shits!' all the time," Richardson said. He rattled some off: impurities that get into piping, temperatures that are too low and end up gumming up the liquid-metal pumping system, a strange sound that can't be traced. He went further, laying out some of the technical challenges that lie ahead. For example, they have to show they can compress the plasma in a way that doesn't compromise its symmetry. Remember the earlier balloon analogy? General Fusion has to prove that its plasma balloon, in the shape of a doughnut, can be compressed to fusion conditions without losing its form — without some part of the plasma bulging out and destroying its original shape.

They also don't really know what happens to the plasma at peak compression. To get some insight, they went cheap and low-tech by turning to high explosives. They trucked a metal-reinforced shipping container into a licensed test field. A small plasma injector inside the container shot plasma into a metal can on the roof. About one kilogram of explosives around the can was detonated, causing the plasma to compress. From that experiment, Laberge and his team were able to collect valuable data. At the time of writing, the company was preparing to use 50 kilograms of explosives for another experiment that would demonstrate net-gain. Why play with explosives in a field? "The big metal sphere in our reactor is going to cost us a lot of money," explained Laberge, who met up with us later in the tour. "But what if the physics don't work? We need to demonstrate the physics beforehand, that way we take financial risk out of building the sphere." Such crude experiments were talked about in the 1970s but never carried out. "I don't know exactly why, but there is evidence in the literature that people planned to do it. They just never did," he said.

GOOD SKILLS HUNTING

Another major challenge has less to do with technology and engineering and everything to do with human resources. Laberge has managed to recruit many former engineering colleagues from Creo, but finding people with a background in plasma and fusion physics — whether seasoned scientists or younger newcomers — has been a frustrating experience for him. "We made job offers to all the guys in the business who are big in the government labs, have lots of experience, and have a little gray in the hair. They all turned us down. Nobody will risk his career in a nice Los Alamos lab to go to a small startup in Canada," said Laberge, adding that recruiting recent post-doctorates in the subject has also been difficult. "They are reluctant to come because it might make a dent in their otherwise awesome résumé. There's this stigma in going to work at a private company in fusion. Very few scientists in this field are willing to go for it." Peter Hagelstein, the MIT professor who continues to push forward on cold fusion research, has drawn attention to this sad state of affairs: the pool of talent willing to carry on with unconventional, outside-the-box fusion research is evaporating. "There are fewer and fewer people each year, the amount of resources is not great, and basically we're getting older — we're not going to live forever," he said. "There's a very good chance we're not going to finish our work before there's none of us left. Generally, new people have not been joining the field . . . If a young person thinks of joining the field, he or she does it at peril of having a career ruined by association."[22]

It's not just conventional fusion scientists thumbing their noses. Even more dismissive is the contingent of scientists and engineers who have devoted their careers to technologies that instead of fusing atoms split them apart. Called nuclear fission, this speciality offers what fusion science doesn't: an industry. There are roughly 440 fission-based nuclear power plants around the world.[23] Collectively they produce about 14 percent of the planet's electricity, and nuclear lobby groups are pushing hard for politicians to embrace a rapid-build nuclear program as part of efforts to reduce greenhouse-gas emissions from power plants. Fission scientists represent the establishment, and for them, fusion power plants are a pipe dream. But the light-water nuclear reactors sold by Westinghouse Electric, General Electric, and France's Areva, as well as Canada's heavy-water CANDU reactors, are still complicated and expensive beasts. The radioactive waste they create is deadly and must be stored safely for hundreds of years. Reactor meltdowns, considered rare, are always a lingering risk. The world was reminded of this disturbing fact on March 11, 2011, when a 9.0 magnitude earthquake and ensuing tsunami pummelled Japan's Fukushima Daiichi Nuclear Power Plant, causing fuel rod meltdowns in three of six reactors. Huge amounts of radiation leaked into the ocean and surrounding air, contaminating fish, seaweed, crops, and other local sources of food. Authorities had no choice but to cull thousands of livestock within the area. More than 80,000 residents within 20 kilometres of the plant, and many beyond 30 kilometres, were forced to evacuate without knowing when, if ever, they could return. It may be years before the full impact of the disaster — on the Japanese economy, on affected communities, on thousands of human lives, and on the environment — is known.

There's also the problem of "dual use" — the fact that the plants and equipment used to enrich uranium as fuel for power reactors can also be used to make highly enriched uranium for a nuclear bomb. The fear is that the more fission-based nuclear power plants we build, the higher risk there is of weapons proliferation. Fusion doesn't carry this heavy risk. A fusion reactor can't melt down, doesn't encourage proliferation, and creates none of the dangerous waste that comes from a fission reactor. Fusion power, if it were to be proven successful, could spell the decline of fission power and the industry built around it.

But fusion power has always been just around the corner, the forever-emerging technology that's going to save the planet. It's the energy breakthrough that keeps crying wolf, or "Eureka!," and over time most people have simply tuned out. I asked Laberge what he considered his odds of success. "I put my chances at 50 percent," he said. "And by the way, I was at 30 percent last year, so it's getting better." Success means different things to different people in different situations. For Laberge, success may be simply proving his doubters wrong by demonstrating that the impossible is possible. Merely showing possibility, however, is a fraction of the fusion journey. Even if it works, the established energy industry will still resist. Regulators, influenced by the old boys' club, will take their time approving it. Utilities, never a group to take a chance on a technology without a track record, will drag their feet. For a General Fusion power plant to have any impact during our lifetimes will require serious backstopping from the government.

Laberge isn't blind to these challenges. He hopes he will see a couple of 300-megawatt beta plants in commercial operation by 2020. After that it may take decades for meaningful adoption. But forecasting is a mug's game. All he can control is his own work — proving that cheap, safe electricity through commercial fusion can be done, sooner rather than later. "I believe if you come with a solution that makes sense, the resources to build your product will come."

It's that kind of thinking that drives innovation, and that drove inventors like Nikola Tesla to forge on despite the many obstacles in their path. But in at least one way Laberge has it easy. He doesn't have to build his power plant 36,000 kilometers above the surface of the Earth. That, as you'll read in the next chapter, is what aerospace veteran Gary Spirnak is attempting to do.

Notes:

1 The first stage of a thermonuclear bomb relies on nuclear fission (the splitting of ura-

nium or plutonium isotopes), which releases energy, creating the heat and compression needed to trigger the second stage fusion reaction of the hydrogen isotopes tritium and deuterium. Fission, therefore, sets off the runaway chain-reaction that leads to fusion. A bomb based strictly on fission is called an atomic bomb.

2 Based on currency rates as of September 9, 2010.

3 Jonathan Leake and Elizabeth Gibney, "Hunting the Holy Grail of Fusion," *The Sunday Times*. September 9, 2007.

4 The National Cold Fusion Institute was shut down after its initial funding ran out.

5 United States Department of Energy, *Cold Fusion Research: A Report of the Energy Research Advisory Board*. November 1989.

6 Sharon Weinberger, "Warming Up to Cold Fusion," *Washington Post*. November 21, 2004.

7 Lisa Zyga, "Physicist claims first real demonstration of cold fusion," Physorg.com. May 27, 2008.

8 "Italian Scientists Claim to Have Demonstrated Cold Fusion," PhysOrg.com. January 20, 2011.

9 Mark Peplow, "Collapsing bubbles have hot plasma core," Nature.com. March 2, 2005.

10 "Bubble fusion: silencing the hype," Nature.com. March 8, 2006. http://www.nature.com/news/2006/060306/full/news060306-1.html. (First article in four-part series.)

11 "Report of the Investigation Committee in the Matter of Dr. Rusi P. Taleyarkhan," Purdue University. April 18, 2008.

12 Eugenie Samuel Reich, "Bubble-fusion scientist debarred from federal funding," Nature.com. November 23, 2009. http://www.nature.com/news/2009/091123/full/news.2009.1103.html.

13 It's an interesting situation, because mainstream fusion scientists are themselves the outcasts of mainstream nuclear science, which is preoccupied mostly with nuclear fission technology. So you can imagine how far "out there" the cold fusion and bubble fusion crowd is viewed, along with ventures like General Fusion.

14 Nikola Tesla was known for creating "ball lightning," which was actually self-confining and quite stable plasma spheres. But no labs have been able to replicate Tesla's plasmoids, which the Serbian-American engineer considered a fascinating nuisance.

15 According to ITER, "Neutral Beam Injectors are used to shoot uncharged high-energy particles in the plasma where, by way of collision, they transfer their energy to the plasma particles."

16 "NIF Megajoule Shot Shatters Record" (news brief), Lawrence Livermore National Laboratory. February 5, 2010.

17 SDTC is also General Fusion's bigger public funder, but not its only one. The company received $412,500 from Canada's National Research Council in 2007.

18 Josh Dean, "This Machine Might Save the World," *Popular Science*. January 2009.

19 Abraham Loeb, "The Right Kind of Risk," *Nature* 467 (September 16, 2010): 358. (Loeb is the director of the Institute for Theory and Computation at Harvard

University.)

20 Because Chrysalix became involved as an investor, Brown and Fletcher had to donate their own personal shares in the company to charity to avoid a conflict of interest.

21 "General Fusion Closes $19.5 Million Series B Funding Round," General Fusion press release. May 5, 2011.

22 "An Interview with Peter Hagelstein," io9.com. March 23, 2010.

23 World Nuclear Association, "Nuclear Power in the World Data," updated to September 2010. www.world-nuclear.org.

CHAPTER 2

Out of This World

Beaming Solar Power from Space

*"The energy of the sun was stored, con-
verted, and utilized directly on a planet-
wide scale. All Earth turned off its burning
coal, its fissioning uranium, and flipped the
switch that connected all of it to a small
station, one mile in diameter, circling the
Earth at half a distance from the moon. All
Earth ran by invisible beams of sunpower."*
— Isaac Asimov, "The Last Question" (1956)

Replicating a fusion process that takes place inside the sun is one
way to bring unlimited amounts of clean energy to humanity.
Doing a better job of harnessing what the sun already does is
another.

Two days after visiting General Fusion, I hopped on a plane
to Los Angeles to visit the Manhattan Beach headquarters of
Solaren Corporation, a tiny startup with supersized ambitions. The

company captured headlines on April 13, 2009, when California's biggest energy utility, Pacific Gas and Electric (PG&E) of San Francisco, announced an unprecedented agreement to purchase all the electricity from a 200-megawatt solar power plant to be designed, assembled, and operated by Solaren. It wouldn't be such big news except for one not-so-trivial fact: instead of being built on a sun-drenched desert, the power plant would be parked in orbit about 36,000 kilometers above the Earth's surface. The world's first space-based power station would harvest energy from the sun, the ultimate nuclear fusion reactor, and use microwaves to beam that energy down to Earth where it would be converted into clean electricity. Solaren plans to build the entire system in California, pack it tightly enough to fit into four, maybe five, rockets, and then launch the pieces into space. Once in orbit the pieces will automatically unfold, use special sensors to find each other, and then be remotely assembled by Solaren engineers sitting safely and comfortably in a Los Angeles control room. No astronauts; no danger pay required. Electricity, remarkably, is expected to start flowing to California homes and businesses by 2016.

Solaren's breakthrough PG&E deal, which was approved months later by California's public utilities commission, attracted a barrage of media attention. But it also drew equal amounts of criticism and questions. Who were these space mavericks? How serious could they be? Was this just a money-making scheme with no chance of success? How could a small private company possibly follow through? Outer space, after all, is the domain of big government.

I have to admit, visiting Solaren's office didn't inspire much confidence. No high-tech campus or fancy lobby with a wall-mounted plasma TV running a slick corporate video. No small-scale models showing how a space-based solar power station would look. In fact, when I arrived I thought I had the wrong address. My taxi dropped me in front of a sprawling building on a well-manicured property surrounded by palm trees. The building was a 12-year-old movie studio called Raleigh Manhattan Beach Studios, built by Walt Disney's nephew Roy and home to Marvel

Studios. That's right, it turned out Solaren's headquarters is also where *Iron Man 2*, *Thor*, and *The Avengers* were filmed, along with more than a dozen prime-time TV shows, including *Boston Legal* and *CSI: Miami*. After walking past a clothes trailer beside the set of *90210*, I eventually found my way to Solaren's office, tucked away on the far end of the building.

Cal Boerman, the company's director of energy services, met me at an elevator bank and took us to the second floor. Solaren had just moved in a few months earlier, he explained, as we walked through 18,000 square feet of bright and empty office space with that telltale new carpet smell. "It was like a cave when we first got it, everything was very dark and dank," said Boerman, who told me that the previous tenant was the editing and prepping offices for *The Biggest Loser*. The office comfortably accommodated about 60 staff, plenty of room for this eight-person operation to grow. But on the day of my visit, the walls were still bare and individual offices appeared unused. A big "Solaren" sign hung in an otherwise empty reception area.

It was a funny juxtaposition: a small startup trying to make

Cal Boerman (left) and Gary Spirnak pose beside a new Solaren sign at the company's new — and at the time very sparse — office in Los Angeles.

what many consider the stuff of science fiction a reality, housed in the same building as a studio trying to make these kinds of impossible-seeming ideas a reality on the big screen. It wasn't until company co-founder and chief executive Gary Spirnak entered the conference room and began laying out Solaren's strategy that I began to think these guys were the real deal — and, well, just might pull it off. Sure, there are those who say it's too risky an enterprise to ever work, or too expensive to justify in the foreseeable future. At least one top NASA official has dismissed the concept as "fiscally ridiculous." Spirnak, who could pass for actor Kevin Spacey's taller and cleaner-cut twin, isn't shaken by such comments. In fact, he more than welcomes the skepticism: the fewer people who follow his path the better. "We would like to be in a position where in two years, as we get this off and running, it will take a consortium of countries to catch us. That's the kind of head start we want," he said. "This is not something that intimidates us. We know exactly what we're doing, and we know how hard it is."

AN UNFOLDING FANTASY

Spirnak's confidence in his bold mission is best put into perspective with a bit of a history lesson. The idea began in the realm of science fiction and the work of writer Isaac Asimov, who is credited as the first person to mention the concept of space-based solar power in his 1941 short story "Reason," as well as in later writings such as "The Last Question." In "Reason" (part of Asimov's *I, Robot* series), two engineers, Gregory Powell and Mike Donovan, work on a station in space that beams energy to Earth and other nearby planets. "Our beams feed these worlds energy drawn from one of those huge incandescent globes that happens to be near us," Powell explains to Cutie, a robot he has built. "We call that globe the sun."[1]

Nearly three decades would pass before reality caught up with fantasy. In 1968, 45-year-old aerospace engineer Peter Glaser first presented the idea of a space-based solar power station at an industry gathering in Boulder, Colorado. His speech was met with raised eyebrows and head-scratching. "Some of the members of the audience thought I was discussing science fiction," recalled Glaser in his unpublished memoirs, parts of which his wife, Eva, kindly shared with me.[2] That same year Glaser published a paper in the research journal *Science* that detailed how one might go about building a system in orbit that collects energy from the sun and beams it to a massive receiving station on Earth. In 1973 Glaser became the first person to obtain a U.S. patent on this design concept.

The Czech-American engineer was far from an unknown name in aerospace circles; he was, in fact, deeply involved in the U.S. space program. In the late 1960s Glaser led a team of engineers who designed a special laser-reflecting system that was deployed on the moon as part of the historic *Apollo* 11 lunar landing mission. Astronaut Buzz Aldrin installed the device, which is a special type of mirror that can reflect an incoming laser beam back to the location from which it came. It's the only experiment on the lunar surface that's still operational today, and it allows scientists to measure the distance between the Earth and the moon, which we now know is expanding by about an inch every year because of the sun's gravitational pull.

Glaser's interest in space-based solar power seems to have evolved out of his early days as an engineer at management consulting firm Arthur D. Little. One of Glaser's first major assignments was to develop a way to expose certain materials to extremely high temperatures, mimicking what a spacecraft would experience upon reentry into the Earth's atmosphere. He designed a solar furnace for his experiments but became frustrated every time a cloud blocked out the sun. That may have planted in his mind, early on, the idea that tapping the sun's energy was most valuable when done beyond the clutter of Earth's atmosphere.

It would be unfair to imply that Glaser's idea of a space-based solar power plant was filed away under a folder labeled "Wacky Concepts." Scientists and engineers outside the field may have been skeptical in the late '60s, but the idea intrigued both the U.S.

Department of Energy and the National Aeronautics and Space Administration (NASA) and over the past three decades about $80 million has been spent studying the concept.[3] Throughout much of the 1970s both agencies conducted thorough design and feasibility studies. In 1981 the U.S. Office of Technology Assessment brought all of these studies together in one 300-page report that warned against rolling the dice on what it termed solar power satellites (SPS). "Too little is currently known about the technical, economic, and environmental aspects of SPS to make a sound decision whether to proceed with its development and deployment," the report concluded. "Even if it were needed and work began now, a commercial SPS is unlikely to be available before 2005–2015 because of many uncertainties and the long lead time needed for testing and demonstrations."[4]

A number of potential barriers were cited. The public, it said, would worry about environmental and health risks, mainly related to the idea of invisible beams of energy striking the Earth. Wouldn't these microwaves cook birds and animals that got caught in their path? High cost was another major obstacle, as well as military concerns. How, after all, could we protect a solar power plant in space from being shot down by the Russians or some other rogue nation? The environmental lobby was also flagged as a potential roadblock because efforts directed at space-based solar power could drain resources away from small-scale terrestrial solar technologies. The latter put power into the hands of individuals and communities, which was considered a good thing. By contrast, space-based solar power — like nuclear power — further concentrated decision-making and control in the hands of large government agencies and utilities. Not such a good thing.

The 1981 report, which came out during President Ronald Reagan's first year in the White House, promptly faded into the background, and serious talk of space-based solar power didn't resurface until the mid-1990s. Reagan slashed the renewable energy programs created under previous president Jimmy Carter; Reagan was more interested in funding the Strategic

Defense Initiative, which came to be known as "Star Wars," a space-based missile system that could defend the United States from a foreign attack.

A FRESH LOOK

NASA revived its interest in space-based solar power early in the Clinton years. In 1995 the agency launched its "Fresh Look Study" to find out if the concept, measured against the latest technologies and projected advances in solar cells, robotics, and launch capabilities, was ready for a big government push. The study concluded that, indeed, the concept was more viable and that more detailed research was needed, sparking a series of research efforts throughout the late 1990s. Consultations with outside technologists and an independent economic and market analysis culminated in the creation of the Solar Power Exploratory Research and Technology (SERT) program.

SERT was enthusiastic about space-based solar power but its findings didn't sugar-coat the challenges that still remained. What it did do was set aggressive but, in its view, achievable timelines for development. SERT concluded that with strong government support it would be possible to have a one-megawatt prototype in orbit by 2011 that would allow for testing of space-to-ground wireless power transmission. Within 15 to 20 years (i.e., 2015 to 2020), it envisioned an "intermediate-scale" solar power plant roughly 10 megawatts in size that could beam power to the Earth or an outpost on the moon. This prototype could also demonstrate its future contribution to space exploration.

Large prototypes in the one- or two-gigawatt category would enter orbit after 2025, SERT projected. These stations would have roughly the same power capacity as a modern nuclear power reactor and, like a nuke plant, could supply this power to Earth 24 hours a day. If we wanted to go even larger we'd have to wait until after 2050, when giant 10-gigawatt solar stations might be deployed on a commercial basis. "Such systems might find application in providing very-large-scale power to terrestrial markets,

for the industrial development of space resources, or in powering robotic probes to near-interstellar space during the latter
portion of this century," explained NASA's John Mankins during
testimony to a U.S. House of Representatives science subcommittee in 2000.[5] (Mankins, it should be noted, is a determined
promoter of space-based solar power. He ended up leaving NASA
and is now promoting the concept through his own consulting
firm, Artemis Innovation.)

It was around this time that Gary Spirnak, then a project engineer with satellite builder Hughes Space and Communications
Company, became preoccupied with the concept of space-based
solar power, the opportunities and the obstacles of making it real.
He'd been aware of the idea since his college days in the late 1970s,
but as a young man he was more interested in playing baseball. A
shoulder injury put a quick end to that, so the Pittsburgh native
decided to finish up his engineering degree and enrol in the U.S.
Air Force, where he was given the chance to complete his master's
degree in mechanical aerospace engineering. It wasn't long
before his career in aerospace took off. In 1986, shortly after the
Challenger shuttle disaster, Spirnak was sent to Cape Canaveral
to oversee intelligence missions and coordinate classified shuttle
flights for the Department of Defense. It wasn't the best of times.
The *Challenger* disaster put a two-year hold on shuttle missions,
and Spirnak was being nudged toward desk jobs that didn't interest him. So he left the air force and in 1991 joined Hughes, first
in its D.C. office but soon after in Los Angeles, where the thirty-
something engineer became project manager for a number of
advanced space initiatives.

For Spirnak, a company like Hughes was a breath of fresh
air. He liked the company's "spunky" culture, a description that
reflected the character, determination, and entrepreneurial spirit
of its famous founder, Howard Hughes. "We used to joke that the
reason for Hughes' success is that nobody was in charge," said
Boerman, who also used to work under the Hughes umbrella.

Spirnak nodded in agreement. "When I was in the air force
if you wanted the impossible done, you went to Hughes. If you

wanted first-of-its-kind new stuff, impossible stuff, you went to these guys." Spirnak enjoyed his time at Hughes, but after 10 years there his bosses suggested he get his MBA to round out his résumé. "I kind of went kicking and screaming because I was having a lot of fun building satellites." It did, however, offer some time for reflection. While completing his MBA program Spirnak, who had just turned 40, began to see that the maverick spirit that made Hughes attractive as an employer was at risk of fading.[6] Military contractor the Boeing Company made a $3.75-billion offer for Hughes' satellite manufacturing business, which in late 2000 ended up merging with Boeing's own aerospace group.

"When Boeing bought Hughes it felt like I was kind of back in the air force," said Spirnak. "Boeing was very much this command-and-control organization. Hughes wasn't like that. It was very independent." Spirnak figured it was time to move on.

Armed with an MBA and engineering degree, as well as experience in the U.S. Air Force and at Hughes, Spirnak found himself being courted into the lucrative consulting market. He had several offers, all of them financially enticing, but his heart was in the satellite field his career was built on. That's when space-based solar power, the concept that had so intrigued him as a college student more than two decades earlier, resurfaced as more than just a boyish curiosity. He approached his long-time colleague and friend Jim Rogers, a former chief scientist at Hughes Aircraft's optical systems division who had become an industry consultant. "I remember telling Jim I was thinking about this idea of space solar power," said Spirnak. "I told him, 'I just think we've worked through tougher problems than this. I bet we could figure this out.'"

Rogers was on the same wavelength and in 2001 the two men agreed to pursue it. They began digging into a quarter century of NASA research and found that many of the obstacles identified in the 1970s and even the 1990s had been, or would soon be, overcome. They also thought that NASA had analyzed space-based solar from the wrong perspective. NASA was hung up on cost, specifically launch costs, and viewed that obstacle through the lens of a U.S. space program working under a limited budget. This made huge investments in the area difficult to justify against what were perceived as higher-priority government programs. As a result, the U.S. government had yet to pull the trigger on space-based solar. Meanwhile, a risk-averse private sector remained on the sidelines waiting for the government to take the lead. Tired of waiting, tired of hearing calls for more study, Spirnak and Rogers took a different approach. Forget about government, and forget about launch costs they had no control over. They looked instead at space-based solar generation as a profit-making business venture, albeit a high-risk one. They asked themselves: how can we lower the costs? And for how much would we need to sell our electricity to make this an economical enterprise?

WEIGHT LOSS PROGRAM

Electricity costs in North America during the 1970s were low, and they remained relatively inexpensive throughout the '80s and '90s. For example, in 1990 the average all-in retail price for electricity in the United States was 6.57 cents per kilowatt-hour.[7] This low cost is one of many reasons why space-based solar power just wouldn't work as a business proposition or as a government-backed venture, particularly the way NASA envisioned it. NASA's vision and the state of technology during the 1970s and '80s imposed other limitations. NASA wanted to start small, initially sacrificing economies of scale, and from there build slowly and cautiously to something larger over a span of several decades. The agency also saw a need to establish a space factory that would serve as a base for the hundreds of astronauts required to physically assemble space-based solar power systems. There was also the fact that, in the late 1970s, the efficiency of solar photovoltaic (PV) cells, even advanced cells used for space applications, were limited to about 10 percent. Just one out of 10 solar electrons going into a cell would come out as electricity. Of that, another half would get lost during wireless transmission from orbit down to Earth. "The whole system they envisioned from sunlight to electricity was maybe 5 percent efficient," said Spirnak.

Spirnak and Rogers thought they could do dramatically better and were encouraged to learn that the high-end solar PV cells used in space applications were expected to reach about 50 percent efficiency by around 2015. In 2010 Boeing's Spectrolab division, the leading supplier of solar cells for satellites, already sold space-ready cells that were 30 percent efficient, and a year earlier the company had announced it had produced cells boasting efficiency of 41.6 percent, which at the time was a world record.[8] As of June 2011, the record was 43.5 percent held by a San Jose, California, company called Solar Junction. But even with these higher and higher cell efficiencies, the business case still didn't work.

The stumbling block was *weight*. Solaren had to figure out a way to dramatically reduce the weight of its system, otherwise it would simply cost too much to launch all its equipment into orbit. More efficient solar cells helped, because it meant fewer cells were needed to generate the same amount of electricity. But that alone wasn't nearly enough. Undeterred, Spirnak and his team took a closer look at earlier NASA designs and found what they considered a flaw in the agency's approach. "They were trying to move electricity around in space. What they saw was a big group of solar panels all connected by wires that would go for kilometers," he explained, adding that the copper wire required could add up to thousands of metric tons. "The wires were like a third of the weight of their system, literally millions of kilograms of wire. So we looked at this and thought, 'This just doesn't make sense.'"

In the space business (any transportation business, really), weight doesn't just matter — it's everything. The more weight you have, the more launch capacity you need. It takes tremendous amounts of energy to launch an object into space, and that energy translates directly into cost. A 2007 report to the director of the Pentagon's National Security Space Office estimated that a single space-based solar power satellite of one gigawatt or more in size would weigh more than 3,000 metric tons (or three million kilograms) and require 120 launches to get it into orbit.[9] Based on similar design assumptions, Solaren's 200-megawatt system would presumably require 24 launches and weigh north of 600,000 kilograms.

No matter how it was sliced, two dozen launches would kill the business case for Solaren. If copper wiring was the problem, then the solution was to design the wiring right out of the system, thought Spirnak. The company decided to tackle this challenge in three ways.

First, instead of moving electricity around in space through heavy wires, a Solaren system would move sunlight around in space and concentrate it on a much smaller area of high-efficiency solar cells using super-lightweight reflective mirrors made from a durable and flexible material such as Mylar.

Second, the company would have to come up with a better way of connecting the solar cells to the power amplifier, a device that converts electricity into the microwaves that are beamed down to Earth.

Typically a high-powered vacuum tube called a magnetron would be picked for that kind of job, but this approach posed problems for Solaren. When an electric current goes into a magnetron about 85 percent of it is turned into electromagnetic energy (microwaves) that travels at the speed of light toward a target on Earth. The problem is that the electricity coming out of a solar cell is high current and low voltage, while magnetrons require low current and extremely high voltages. This mismatch between current and voltage would require expensive and heavy power electronics to bring them in line.

One alternative to magnetrons that was invented in the 1970s is a solid-state amplifier. Like solar cells, these amplifiers are made from solid semiconductor materials instead of vacuum tubes. These devices match up nicely to solar-cell voltages and current but earlier designs were only 20 percent efficient, meaning 80 percent of the energy that went into them was thrown away as heat. "Over the past few years, however, there have been some breakthroughs with these solid-state devices, and they're now up over 80 percent and even 90 percent efficient," said Spirnak. "So now you can have a solid-state device like a solar cell going to a solid-state amplifier, and you can have matching voltages." In other words, high efficiencies were preserved but there was no longer a need for power electronics and associated wiring. "It just gets rid of all that weight and garbage that went along with the old design," Spirnak added.

Finally, major pieces of Solaren's in-space system would never be physically connected to each other. They would float freely in orbit, some a kilometer or more apart, kept in alignment using a combination of sensors and thrusters. This approach would cut even more weight out of the system, allowing Solaren to get all the pieces in place with four or five launches — a payload totaling about 125,000 kilograms, or less than a quarter of the U.S. National Security Space Office estimate. The company is already in talks with manufacturers Lockheed and Boeing about using the largest available evolved expendable launch vehicles, or EELV Heavies, for the mission. The EELV Heavies are a new class of launch vehicle developed to lower launch costs by 25 percent, and the first "Heavy" was launched for the U.S. Air Force in 2004. Solaren will likely need six in total, including one or two for test launches. Powerful enough to drag five adult elephants into space, they cost "a few hundred million bucks a piece," said Spirnak. "They also take about three years to build from the purchase contract to getting them launched." To meet its obligations to PG&E, that timeline meant that Solaren had to get in the launch queue in early 2011.

One positive development came in March 2011. Space Exploration Technologies, the private rocket maker founded by Tesla Motors chief executive Elon Musk, announced plans to build and launch the world's largest rocket sometime in 2014. Called the Falcon Heavy, it will be able to carry twice as much weight as the next largest EELV rocket, the Delta IV Heavy, but at one third of the cost — or so SpaceX claims.[10] "We are glad to see SpaceX doing this," Boerman told me. "The larger lift capacity helps, because we can put more into each individual launch and maybe eliminate a launch, which saves money and time. So this is going in the right direction for Solaren's needs." Indeed, it could mean as few as three launches to do what NASA thought could only be done with 24 launches.

A BETTER MOUSETRAP

Solaren's new model also meant that deployment would be largely automatic with remote assistance from a control center in California. There would be no need to send astronauts up, and no need for an assembly factory. "We looked at all this, ran the numbers, and thought, 'Wow! It works.' We came up with some good economics numbers," said Spirnak, adding that the system, taking all energy losses into account, will be about 25 percent efficient at taking sunlight from space and turning it into grid-ready electricity on Earth. That's five times more efficient than earlier NASA designs. Spirnak added with a just a hint of cockiness, "We came up with a better mousetrap."

There was also icing on the cake: over the years that Spirnak and his team worked on their design, electricity prices started to shoot up. In 2000 the average U.S. retail price stood at 6.81 cents per kilowatt-hour, but in the years that followed it began to ascend.[11] Indeed, by the end of 2010 average retail prices had rocketed to nearly 10 cents, an increase driven largely by the cost of overhauling aging infrastructure and state government mandates to add costlier green energy to the power mix. Californians, who paid an average retail rate of 8.84 cents in 1990, paid closer to 13.22 cents in 2010 with residential customers paying on average more than 15 cents. In fact, if you live in San Francisco, the headquarters for Pacific Gas and Electric, your total bill breaks down to well over 20 cents per kilowatt-hour. That means leaving a standard 100-watt light bulb on for just over two days would cost you a dollar — more than the price of the light bulb!

This trend worked out well for Solaren. The power-purchase contract with PG&E gave the company a guaranteed rate, somewhere around 13 cents per kilowatt-hour. With that price, as well as a guaranteed revenue stream for 15 years, Solaren could squeeze out enough of a profit from its initial prototype system to satisfy existing investors and lure new ones aboard. Spirnak and Rogers raised about $5 million from friends, family, and small investors during their first eight years to cover the cost of designing their system and filing patents. To finish the 200-megawatt prototype, however, Solaren will need to raise a few billion dollars — the kind of cash required for a new nuclear reactor that has six or seven times the power capacity.

What will Solaren's prototype look like and how will it operate when complete? The general design consists of five main pieces in space and one on the ground. The parts in orbit "are bigger than anything ever flown in space before," said Spirnak.

Imagine a giant inflatable salad bowl made of ultrathin Mylar that's about one kilometer across and is reflective on the inside. Solaren is careful not to reveal too much information for proprietary reasons, but that's the basic idea. That gigantic salad bowl, call it the primary mirror, would always face the sun. It would be packed tightly inside one EELV Heavy and, once launched and positioned in its proper orbit, would automatically open up like an umbrella. The primary mirror's job is to direct sunlight to a much smaller secondary mirror that would focus the solar energy before reflecting it onto the surface of a separate generator module composed of tens of millions of tiny solar cells. The surface of this module will likely cover an area of half a square kilometer — maybe less depending on solar-cell efficiency. The solar cells generate direct-current electricity that flows to a neighboring solid-state power amplifier, which converts the electricity into microwaves. Finally, those microwaves are bounced off a reflecting antenna shaped like a massive satellite dish that, like the primary mirror, is about a kilometer wide. The big dish beams the microwave energy down to a sprawling receiving station located in central California.

Solaren plans to locate that receiving station in Fresno on fallow land less than two kilometers from PG&E's closest

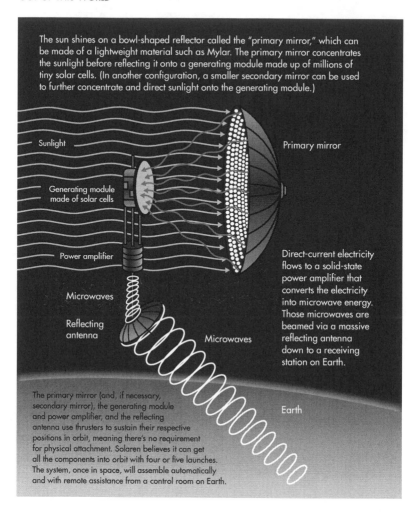

The sun shines on a bowl-shaped reflector called the "primary mirror," which can be made of a lightweight material such as Mylar. The primary mirror concentrates the sunlight before reflecting it onto a generating module made up of millions of tiny solar cells. (In another configuration, a smaller secondary mirror can be used to further concentrate and direct sunlight onto the generating module.)

Sunlight

Primary mirror

Generating module made of solar cells

Power amplifier

Microwaves

Reflecting antenna

Microwaves

Direct-current electricity flows to a solid-state power amplifier that converts the electricity into microwave energy. Those microwaves are beamed via a massive reflecting antenna down to a receiving station on Earth.

The primary mirror (and, if necessary, secondary mirror), the generating module and power amplifier, and the reflecting antenna use thrusters to sustain their respective positions in orbit, meaning there's no requirement for physical attachment. Solaren believes it can get all the components into orbit with four or five launches. The system, once in space, will assemble automatically and with remote assistance from a control room on Earth.

Earth

transmission line. The receiving station is composed of thousands of devices called rectifying antennas, or rectennas, which are tilted toward the sky like your typical sun-worshipping solar panel. The rectennas take the microwave energy beamed from space and turn it back into electricity that is sold into PG&E's electrical grid. The kilometer-diameter microwave beam becomes more than twice that wide by the time it reaches the Earth's surface, so the receiving station must cover an area of more than

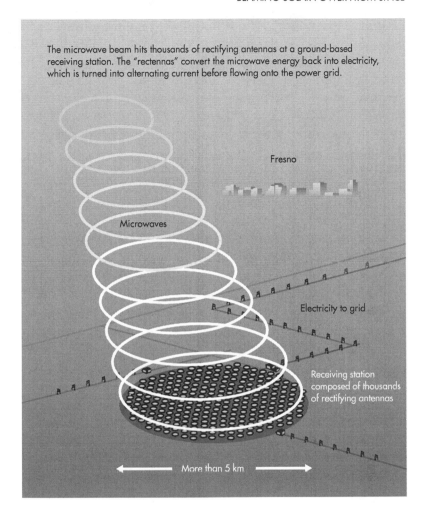

The microwave beam hits thousands of rectifying antennas at a ground-based receiving station. The "rectennas" convert the microwave energy back into electricity, which is turned into alternating current before flowing onto the power grid.

Fresno

Microwaves

Electricity to grid

Receiving station composed of thousands of rectifying antennas

More than 5 km

five square kilometers. It may seem like a large target, but from 36,000 kilometers away it's akin to throwing a bull's-eye on a dart board from 50 meters away.

CUTTING THE CORD

The idea of transmitting power through the open air has fascinated people for decades.[12] In fact, it was radio inventor Nikola

Tesla who first saw it was possible to not only move voices and information through the atmosphere over vast distances, but also electricity. With an investment from financier J.P. Morgan, Tesla famously began building Wardenclyffe Tower — known also as the Tesla Tower — in 1901 in Long Island, about 100 kilometers from New York City. The wood-supported tower was nearly 200 feet tall and had a large metal transmitter on the top in a half-sphere shape. It was used by Tesla in those early days to experiment with wireless communications *and* power transmission. In 1908 he described how Wardenclyffe would usher in a future without wires:

> As soon as completed, it will be possible for a business man in New York to dictate instructions, and have them instantly appear in type at his office in London or elsewhere. He will be able to call up, from his desk, and talk to any telephone subscriber on the globe, without any change whatever in existing equipment. An inexpensive instrument, not bigger than a watch, will enable its bearer to hear anywhere, on sea or land, music or song, the speech of a political leader, the address of an eminent man of science, or the sermon of an eloquent clergyman, delivered in some other place, however distant. In the same manner any picture, character, drawing, or print can be transferred from one to another place. Millions of such instruments can be operated from but one plant of this kind. *More important than all of this, however, will be the transmission of power, without wires, which will be shown on a scale large enough to carry conviction.*[13] [emphasis added]

Tesla's description of a world without wires will sound familiar to BlackBerry-addicted executives and iPhone-toting consumers, who can today transfer pictures to each other and download movies from the Internet without any need to plug in. Few would have expected such advancements three decades ago, let alone in Tesla's day. Microsoft chief executive Steve Ballmer put it this way when speaking in early 2010 at one of the world's largest consumer electronics conferences, "The things we take for granted now would have sounded like science fiction in the early 1980s."[14] Until recently the only wire left to fall was the power cord, but those who follow the latest technology trends know that wireless charging of gadgets over short distances is quickly becoming a reality.

BEAM ME DOWN, SCOTTY!

But Tesla, who did experiments with short distances, also had a much grander vision. He believed air was a conductor, just like any conducting metal, capable of transmitting the energy equivalent of thousands of horsepower over hundreds, even thousands, of kilometers.[15] "Its practical consummation would mean that energy would be available for the uses of man at any point of the globe," he wrote in 1900.[16] In 1927 he talked about the possibility of sharing the great hydroelectric potential of Niagara Falls by beaming the electricity it generated through the upper atmosphere to faraway places.[17] Tesla's dream was to build what he called his "World System" — a wireless transmission infrastructure that carried voices, information, *and* power. On September 3, 1911, the *New York American* newspaper published an illustration that shows Tesla's World System providing electricity through the air to power automobiles, trains, ships, and airplanes.

As far as voices and information go, Tesla's vision proved remarkably accurate. The same can't be said — yet — for transferring bulk power over long distances, but Telsa's work did inspire several generations of scientists and engineers. American engineer William Brown, who pioneered the concept of microwave power transmission during the 1960s, has cited Tesla as a source of inspiration. Brown is credited with inventing the rectenna, the same basic device that Solaren plans to deploy in the thousands at its receiving station in Fresno. Brown captured the imagination of Americans in 1964 by demonstrating the technology to the national press, including media big shot Walter Cronkite.

OUT OF THIS WORLD

Brown flew a small model of a helicopter using a microwave beam as its sole source of power. Unfortunately the U.S. Air Force, the prime target for Brown's research, wasn't interested enough to fund further work, and Brown was left to finance his own research during the late 1960s. It was only after Peter Glaser introduced his concept of space-based solar power that NASA and the U.S. Department of Energy, in the late 1970s, saw the potential of microwave power beaming and began funding further research. "The technology has now reached a high level of maturity and represents an available and possibly invaluable resource for immediate and future use," Brown wrote in 1984.[18]

If Brown were alive today (he died in 1999), he would likely be disappointed with how little the technology has been embraced in the intervening 25 years. Certainly experimentation has continued, with the United States, Canada, France, Japan, and Russia (and the former Soviet Union) taking the lead. Canada's stationary high-altitude relay platform (SHARP) program is one example. During the 1980s government scientists built a pilotless surveillance plane that could be powered and remotely controlled using microwaves beamed from the ground. The plane had its maiden flight on September 17, 1987, and flew for 20 minutes. Three weeks later they kept it flying for more than an hour. Canada eventually discontinued the SHARP program, but the Japanese conducted similar flight tests in 1992.[19]

There's no question that recent interest in space-based solar power has sparked new appreciation for the potential of microwave power beaming and the need to demonstrate it at greater distances. One effort in 2008 beamed power 150 kilometers from a mountaintop in Maui to a group of receiving antennas on the Hawaii main island. The efficiency of the transfer was terrible, and only 20 watts made it through, but for the U.S. and Japanese scientists who set it up it was an unprecedented success.[20] I got to see a demonstration of this technology myself during the International Symposium on Solar Energy From Space, a three-day conference held in Toronto in summer 2009. Nobuyuki Kaya, vice dean of graduate engineering at Kobe University in

Japan, aimed a microwave beam about 10 meters across a large auditorium to power a simple robot.

At the same conference, I met aerospace engineer Kieran Carroll, vice president of technology at SPACE Canada. Carroll told me there was huge potential for microwave power beaming, not just from space to Earth but also between two points on Earth. Take the Canadian province of Newfoundland and Labrador, which has plenty of hydroelectric potential on Labrador but lacks the transmission infrastructure to bring it to the island of Newfoundland. One problem of building that infrastructure is the Strait of Belle Isle, which is 18 kilometers across at its narrowest point. Laying transmission cables on the seafloor would be expensive and risky, partly because icebergs fill the strait every winter and drag the bottom. Carroll has proposed beaming 1,000 megawatts of hydroelectric power from Labrador up to 30 kilometers across the strait to rectenna locations on Newfoundland. To do this successfully on Earth, Carroll told me, would be an ideal way to prove that the technology will also work from space. So far, the idea has fallen flat with provincial and industry authorities, which in November 2010 announced a plan to use submarine cables.

I asked Spirnak about the fact that there has never been a demonstration of microwave power transmission from space to the Earth's surface. He quickly corrected me. "At a small kilowatt level satellites do this every day," he said. "They have solar arrays. They collect sunlight. They use solid-state amplifying and wave tubes. The difference is [satellite providers] throw away the energy and keep the content. But it's the exact same thing." He pointed to DirecTV, the direct-broadcast satellite service founded by Hughes Electronics and launched in 1994. Now controlled by media conglomerate Liberty Media, DirecTV transmits digital television and audio signals to the United States and Central America with a fleet of 10 satellites, each positioned in its own geostationary orbit.

Spirnak explained that each DirecTV satellite has a transmitting antenna about two meters wide that sends a relatively weak microwave beam to Earth. By the time the beam reaches our planet's surface, it blankets about five million square kilometers, or roughly half the area of the United States. The microwaves, which are generated by the solar panels on the satellite, need only be strong enough to act as a carrier of content — that is, the microwaves are an invisible highway for transmitting TV and audio to DirecTV's 18 million subscribers. "The satellite providers don't care about the energy; they just want the content," said Spirnak. If each subscriber has a receiving dish that's about two square feet that means all DirecTV dishes together cover an area of about 36 million square feet. But since the satellite's microwave beam blankets five million square kilometers, or the equivalent of 54 trillion square feet, it means 99.999999 percent of the energy beam is wasted.

What makes Solaren different from DirecTV's setup is that it has a transmitting antenna that's one kilometer in diameter instead of two meters, and it has more than 200 megawatts of solar power feeding it instead of just a few kilowatts. It also has rectennas on the ground waiting to convert that microwave beam to electricity. "So what we're doing is focusing the energy, because we have a much larger aperture in space," explained Spirnak. "We could do that with a DirecTV satellite, and you'd have all that energy going down in two square miles. Now, they wouldn't be very happy about only covering two square miles, but hell, you'd get a really great TV signal."

SPACED OUT

Solaren still does plan to conduct its own microwave power beaming tests, first by transmitting a few kilowatts in their office laboratory and after that in the desert, where they'll beam a megawatt of power across a distance of one or two kilometers. The company will also conduct a beaming test from space, likely in 2014, before proceeding with the pilot power plant. Though some testing is needed, the company dismisses NASA's notion that you have to start with a small power plant in space and gradually build up from there. "Space solar power plants, without a doubt, are animals of scale," said Spirnak. A one-megawatt system simply won't generate enough revenue from electricity sales to pay for the cost of the rockets to get it there. "It's the same reason you wouldn't build a one-megawatt nuclear reactor. Regardless of size, there's a certain sunk cost you have."

But talk, as the old saying goes, is cheap. Getting the entire system built, tested, launched into space, positioned in orbit, assembled, and operational by 2016 is what the most optimistic engineer might call mission impossible. Japan, one of the few nations formally pursuing space-based solar power, doesn't expect to have a working space station for at least two decades.[21] Recall that in 2000 the SERT program said it *may* be possible to get a 10-megawatt prototype into space between 2015 and 2020, and that would be with the full, decade-long backing of government and the resources of NASA. A small private company trying to get a solar power station in space and operational by 2016 — a station that's 20 times larger — well, that's bound to attract widespread skepticism.

Pete Worden, center director for NASA's Ames Research Center, is among the skeptical — not specifically of Solaren's claims, but of the entire concept. He calls himself a "reformed zealot," someone who was once a believer and devotee of space based solar power but after years of study and contemplation is now one of its biggest critics. It doesn't make sense, in his view, because it doesn't pass the economic test. It's overly complicated. It's risky. "It's fiscally ridiculous," Worden said during an interview in 2009 on *The Space Show*, an online current-events show for space enthusiasts and professionals.[22] "I really see no possibility for the foreseeable future that space solar power makes any economic sense whatsoever for supplying power to the planet," he told his interviewer. "To my friends and colleagues who are advocates of this, I'm sorry, but Santa Claus is a long, long way from this one." Worden's biggest problem with space-based solar power is that, when stacked up against other alternatives, primarily ground-based solar power generation, it's simply too risky and expensive to make a business case. His evidence? Silicon Valley investors aren't lining up to support it. "We talk to entrepreneurs about this area, ones that have real money, like venture capitalists, and you'll find they're not really interested," he said.

He's got a point: why go to outer space when solar has a bright future right here on Mother Earth? There's no denying the role that traditional rooftop and ground-based solar power systems will play over the coming years and decades as countries around the world transition away from fossil fuels toward a low-carbon energy mix. After all, it was Thomas Edison who famously said to Henry Ford in 1931: "I'd put my money on the sun and solar energy. What a source of power! I hope we don't have to wait until oil and coal run out before we tackle that." Advancements in this area have been stunning, and the cost of solar continues to fall to that sweet spot called "grid parity," which is the point where solar power becomes competitive with the mix of conventional power sources on the electrical grid, mainly coal, natural gas, and nuclear.

The most popular form of solar power generation is photo-voltaics (PV), which can be made from a number of different semiconductor materials. When the sunlight strikes a solar PV cell, it excites electrons in the semiconductor material and creates an electric current. Research in the area of solar PV has focused mainly on improving efficiency — that is, getting the cells to convert more of the sun's energy into electricity — and on cost reduction. This has resulted in dozens of different flavors of solar PV cells and techniques for manufacturing them. Most today are made of silicon wafers, while others are made of thin films of silicon (made by companies such as Applied Materials and Oerlikon) and more exotic and toxic combinations such as cadmium telluride (made by First Solar and Abound Solar) and copper indium gallium selenide (made by Solyndra and Miasolé). The advantage of these thin-film cells is that they reduce cost by requiring dramatically less semiconductor material. For example, a crystalline silicon wafer is about 350 micrometers thick — equivalent to three or four human hairs lying on top of each other — while a thin film on a cell is about 1 micron, or 100 times thinner than a single human hair. And because the film can be chemically deposited on a flexible substrate, like plastic, the cells can be manufactured on a roll-to-roll process similar to how newspapers are printed, allowing for low-cost, high-volume production.

The trade-off is that thin-film cells are only 6 to 9 percent efficient, compared to efficiencies of between 15 and 25 percent for cells made of crystalline silicon wafers. If you want to install solar panels on limited rooftop space, you're likely to choose a more expensive crystalline silicon panel from a company such as SunPower or Sharp. That's because the higher efficiency allows you to draw more electricity from a smaller area. (In space, super-expensive yet highly efficient PV cells made of gallium arsenide are used to reduce weight, and hence launch costs. Makers of these cells include Spectrolab and Cyrium Technologies.) On the other hand, if you own many hectares of brownfield property and space isn't a limiting factor, you're likely to use an inexpensive thin-film solar panel made from a company such as First Solar, which is now manufacturing its solar modules for less than $1 per watt, down from $3 per watt in 2004. At the time of writing, First Solar announced completion of the world's largest solar PV facility, an 80-megawatt facility based in Sarnia, Ontario, not exactly the sunshine capital of the planet.[23]

Research in this area continues to amaze; take southern Ontario, where I live, for example. Scientists at McMaster University in Hamilton, Ontario, are growing silicon nanowires that, while invisible to the human eye, stand up on a substrate, like bristles on a densely packed hairbrush.[24] The idea here is that the sun can strike a larger surface area on a cell covered in nanowires, and more electricity can be produced per square inch. Others are working on ways of using fewer solar cells by concentrating sunlight with special lenses and curved mirrors. A company in Toronto called Morgan Solar has developed an inexpensive concentrating optic that could lead to solar modules that cost the same as thin-film modules but offer three to four times the efficiency.[25] Efforts are also being made to harness more of the sun's energy by using so-called multi-junction solar cells, which contain two or three layers of materials that can capture different parts of the color spectrum in light. Silicon, for example, only captures a small part of the spectrum and throws away the rest as heat. By layering on other materials, such as gallium arsenide, more of the sun's electromagnetic radiation can be harnessed. Ottawa-based Cyrium Technologies does this by using nanoparticles called "quantum dots," which can absorb different parts of the color spectrum, such as infrared light, depending on the dots' size. Organics dyes, solar "light pipes," and solar products that supply both electricity and heat from the same device are among the many innovations driving the market forward.

Here on the ground the solar market is well established and growing stronger. The International Energy Agency says electricity harnessed from the sun could represent up to a quarter of the world's total electricity production by 2050.[26] Half of that solar electricity would come from PV systems installed on building rooftops or large expanses of land. The other half would come from sprawling concentrating solar power (CSP) systems that use the sun's heat to create steam for generating electricity. This is typically done by using sun-tracking mirrors called heliostats to focus the sun's light, creating high enough temperatures to create steam. At that point, a CSP plant works like any other thermal plant, be it coal or nuclear. An advantage with CSP systems is that the heat they generate can be stored (in tanks of molten salt, for example) and retrieved later when electricity is needed. All of this innovation and acceptance of solar does call into question the need to launch these systems into space.

THE ORBITAL ADVANTAGE

Spirnak is quick to point out that solar innovation benefits his project just as much as any project built on the ground. "If somebody comes up with a better solar cell, we'll just plug it into our system," he said, adding that Solaren's system is basically just a clever integration of existing technologies that can be swapped in and out as they improve.

The challenge with solar technology is that no matter how far it advances and how efficient it becomes, it will still have limitations when used on Earth. Nighttime. Clouds. Atmospheric pollution. When the sun doesn't shine, you can't produce electricity, and when the skies are filled with particles that block sunlight, the efficiency of a solar power system is reduced. In 2008 the United Nations Environment Programme warned of a phenomenon called "atmospheric brown clouds," a toxic mixture of soot, sulphates, and aerosols that are dimming the skies of countries such as China and reducing their ability to produce electricity from solar panels.[27] Geography and local climate also determine the business case for ground-based solar. Install a solar PV system in London or Moscow and you'll produce half as much electricity as a similar installation in Los Angeles, Cape Town, or Cairo.[28]

Space-based solar power overcomes these limitations. There is no night in space. No clouds. No brown haze blocking out the sun. Solaren's system, like all space-based solar power designs, would provide a nearly constant, uninterrupted supply of power like any nuclear, hydroelectric, or coal plant. It would face the sun 24 hours a day and be capable of beaming a predictable stream of energy, and it wouldn't be limited by where that stream could go. To London or Cairo; to the tar sands of northern Alberta or a desert in California; to a military base in Afghanistan or a base station on the moon — the result would be largely the same. No different, really, than the way broadcast satellites can transit television signals to any spot on the planet. Solaren's receiving stations would also take up 95 percent less space than a 200-megawatt solar PV plant built on land.

When one looks at the great lengths people are prepared to go to tap solar power on Earth, the idea of going to space doesn't seem so ambitious. One example is the DESERTEC concept, a plan to build solar PV and concentrating solar power systems on 17,000 square kilometers of the Sahara Desert. The project, which has backing from industry giants Siemens, ABB, Deutsche Bank, and Munich Re, would involve building a massive grid of high-voltage DC transmission lines connecting the Sahara to countries in Europe and Northern Africa. The price tag: $550 billion U.S., based on a goal of supplying 15 percent of Europe's electricity demand by 2050. One might be tempted to criticize Solaren's project because of the vulnerability of its solar power station to space debris and sabotage, but is the DESERTEC project any better off? The equipment it plans to use will face potentially damaging sand storms, and placing the heart of the project in an unstable region like Northern Africa does little to alleviate the energy security concerns brought about by the world's overdependence on Middle East oil.

Still, the idea of outer space, let alone beaming energy from space, overwhelms many people. Jonathan Coopersmith, a technology historian at Texas A&M University, has said that space-based solar power suffers from two things: the giggle factor — "you've got to be joking" — and the Death Star factor — the belief that beaming energy from space is dangerous. A reference to the giant weaponized sphere in *Star Wars*, the Death Star factor could prove to be Solaren's biggest challenge in an age where some people believe even wind turbines are a health risk.[29] Solaren's microwave beam will have some heating effect and the receiving station will be off limits, but if a person happened to hop a security fence and run to the core of the beam, she could probably be there for a few minutes before exceeding exposure standards. The company says planes and birds can safely fly through the beam, though the heat may temporarily draw some animals to it. Solaren will have to go through an environmental assessment before getting the go-ahead, but given that nuclear power plants are far more complex and dangerous, the company doesn't anticipate any deal-breakers. "This is not a giant bug zapper," said Spirnak. He would be naive, however, to not anticipate some public resistance.

The concept is equally overwhelming for investors. For one, most who see the potential of Solaren's approach still want a quick return on their investment. "That just doesn't work for us. We're in it for the long haul," said Spirnak. Others are dismissive from the start. He told me about a meeting he had several years earlier with a potential investor at Duke Capital, a subsidiary of utility Duke Energy. "I was going through my charts with a short presentation and after the second or third chart, the guy says, 'Gary, you've got a typo on your chart.' I asked what it was, and he said, 'Well, your power plant is in space.' I told him he was correct, and he replied, 'We don't do things in space.' That was the end of the meeting." Solaren has run into the same difficulties as General Fusion. Despite providing hundreds of pages of business plans and drawings to investment banks, both big and small, most never took the time to study them. Added Spirnak, "They just don't have a feel for it. Most of the financial community works on this follow-the-leader kind of thing. They're looking for validation somewhere, so if you're coming in as a first-of-a-kind, they're just not willing to take a shot at it."

It doesn't help that one of Solaren's patents talks about the potential of using its microwave beams to bust up hurricanes and modify the weather, an image that conjures up the mad scientist Tesla and his destructive death ray.[30] In theory, the heat from the microwave energy could be used to disrupt air flows that allow hurricanes to form. Spirnak, apparently, wanted to cover all his bases and included that possible use in a patent application. One can only hope that we never have to resort to such risky geoengineering experiments.

For other people more familiar with space exploration and technology, even those who support the pursuit of space-based solar, Solaren is viewed as a black sheep that stands apart from the flock. Part of the reason is that Spirnak doesn't like to attend industry conferences or share too much information with other scientists and engineers interested in the concept, even when asked. That approach has ruffled feathers, yes, but Spirnak is unapologetic. He's not prepared to lose control over intellectual property just to satisfy the curiosity of conference goers, and he has no interest in defending Solaren's plan to armchair skeptics in the crowd. "I've done my master's thesis defense once, I'm not going to do it again," he joked. "People tell us we're hiding something. But this is competitive. It's the same way Toyota doesn't go to Detroit and brief General Motors on what cars it's going to come out with next year." In a way, it's a sign that space-based solar is maturing beyond a cottage industry to a more commercial enterprise. No longer is it just an academic exercise for government scientists living off NASA funding. "At the end of the day, they don't lose their job if they don't make this work," said Spirnak.

UPWARD MOMENTUM

Solaren may have an impressive design for its space system and the chutzpah to pursue its development, but it also has an enormous to-do list. It has managed to attract some strategic investors with patient capital, but it will need to pull in billions of dollars more in investment to turn its paper-based plan into a space-based power station. Solaren needs to order heavy-lift launch vehicles, book time in the launch queue, secure an orbital slot, and, at the same time, build, test, and demonstrate the parts that will complete the puzzle. There are concerned citizens and environmental groups who need convincing that its microwave beam won't kill birds, knock down planes, or cause health issues for nearby communities. It will have to prove that the clean energy from its solar power station will far exceed the energy in the rocket fuel that will launch it into space, and it will have to show investors how it plans to protect the system from space debris and its electronics from solar storms, and how it will guard against and respond to component malfunctions. Solaren will also have to explain what happens in 20 or 30 years when its space station reaches the end of its useful life and risks becoming part of a growing pile of orbiting space junk. It has to do all of this before 2016.

As daunting as all this seems — and is — perhaps Solaren is in the right place at the right time. There's unprecedented chatter out there about the potential of space-based solar power; the Japanese are trying to spark a solar space race; and a number of other private ventures have joined in, including Space Island

Group, PowerSat, and Space Energy Incorporated. Electricity prices are climbing, and power infrastructure is in need of a massive overhaul. Meanwhile, climate change and the environmental consequences of our increasingly treacherous pursuit of oil have a larger percentage of the population looking for cleaner, safer, and abundant energy alternatives. The BP oil spill, the worst in history, offered plenty of motivation for change. It was April 20, 2010, when an explosion on the Deepwater Horizon drill rig in the Gulf of Mexico killed 11 people and caused oil to start gushing from the sea floor at a rate of more than 50,000 barrels a day for nearly three months. Shortly after the spill, Buzz Aldrin, the second man to walk on the moon, issued a call to action. "The timing of the oil catastrophe," he said, "is a great opportunity for reevaluating solar energy from space."[31]

Beyond producing green energy and helping to wean ourselves from oil, Spirnak is just as keen to put infrastructure in space that will open up space exploration, pave the way for missions to distant planets, and create more space-industry jobs. "That drives us as much as anything," he said, careful to give credit where credit is due. "There are a number of things that happened through no fault of our own. Had the U.S. Air Force and NASA not pumped gobs of money into solar-cell research over the last 10 years I wouldn't be here talking to you. Same thing had the air force not gone in after the *Challenger* disaster and built up its unmanned heavy lift capacity."

To not seize these developments, to not run with them when others seem content to walk, to not advance our access to clean energy and space, all simply to avoid the risk of failure — where's the satisfaction in that? Peter Glaser invented the concept more than four decades ago, and he's still waiting for his dream to become reality. In his unpublished memoirs, Glaser lamented what he considers the self-imposed limitations of humankind: "Our inability to see the future except as the continuation of the present is the reason that we let our imagination be bound by commonly accepted perception, despite all evidence pointing to an acceleration of change."

But just because we *can* do something, does it mean we should? We'll explore this question in the next chapter where we'll meet Louis Michaud, the Tornado Man of Sarnia.

Notes:

1 Isaac Asimov, *I, Robot* (New York: Bantam Spectra Books, 2004), 48. (Originally published in 1950.)

2 Retired and living in Lexington, Massachusetts, Peter Glaser was 87 years old when I wrote this chapter. His wife, a sweet woman named Eva, said her husband has early Alzheimer's and is unable to recall much detail of those earlier days.

3 "Space-based solar power as an opportunity for strategic security: Phase 0 architecture feasibility study" (report to the director of the U.S. National Security Space Office), October 10, 2007.

4 U.S. Office of Technology, "Solar Power Satellites." August 1981. 3.

5 John C. Mankins, manager of Advanced Concepts Studies with NASA's Office of Space Flight, testimony to the Subcommittee on Space and Aeronautics, part of the House of Representative's Committee on Science. September 7, 2000.

6 What is it about turning 40 that makes some men reconsider their direction in life and take on risky ventures? You'll recall this was the age that launched General Fusion's Michel Laberge in a completely new direction in life and career.

7 Energy Information Administration (U.S. Department of Energy), "Historical 1990 through Current Month Retail Sales, Revenues, and Average Retail Price of Electricity by State and Sector." http://www.eia.doe.gov.

8 "Boeing Subsidiary Spectrolab Achieves World Record Solar Cell Efficiency," press release. August 26, 2009.

9 "Space-based solar power . . . ," 31.

10 "SpaceX Announces Launch Date for the World's Most Powerful Rocket," SpaceX press release. March 14, 2011.

11 Energy Information Administration (U.S. Department of Energy), "Historical 1990 . . ."

12 For a great overview of different wireless power transfer technology, see "Wireless Power Transfer Technology," Electric Power Research Institute, December 23, 2009. Available for free download at http://my.epri.com.

13 Walter W. Massie and Charles R. Underhill, "The Future of the Wireless Art," *Wireless Telegraphy & Telephone* (1908): 67–71.

14 Steve Ballmer, keynote speech, CES Conference in Las Vegas. January 6, 2010.

15 Some go further than Tesla, suggesting that the upper atmosphere isn't just a conductor but also stores energy that could be harnessed. For a theory on how the Earth and its atmosphere behave as a big giant capacitor that stores energy from solar wind, read chapter 6.

16 Nikola Tesla, "The problem of increasing human energy," *Century Illustrated* (June 1900).

17 Nikola Tesla, "World Systems of Wireless Transmission of Energy," *Telegraph and Telegraph Age* (October 16, 1927).

18 William C. Brown, "The History of Power Transmission by Radio Waves," *IEEE Transactions on Microwave Theory and Techniques* MMT-32, no. 9 (September 1984): 1230–1242.

19 George Jull, essay on history of the SHARP program, http://www.friendsofcrc.ca. (Jull is a researcher at Canada's old Communications Research Centre.)

20 Loretta Hidalgo Whitesides, "Researchers Beam 'Space' Solar Power in Hawaii," Wired.com. September 12, 2008. http://www.wired.com/wired-science/2008/09/visionary-beams/.

21 Shigeru Sato and Yuji Okada, "Mitsubishi, IHI to Join $21 Bln Space Solar Project," Bloomberg News. August 31, 2009.

22 *The Space Show*, March 23, 2009. http://www.thespaceshow.com. Host Dr. David Livingston interviews Dr. S. Pete Worden, center director for the NASA Ames Research Center, on space-based solar power.

23 "Enbridge and First Solar Complete the Largest Photovoltaic Facility in the World," press release. October 4, 2010.

24 Tyler Hamilton, "Flexible, Nanowire Solar Cells," *Technology Review*. February 6, 2008. http://www.technologyreview.com/energy/20163/.

25 Tyler Hamilton, "A Cheaper Solar Concentrating," published online at MIT's *Technology Review*, February 20, 2009. http://www.technologyreview.com/energy/22204/.

26 "IEA sees great potential for solar, providing up to quarter of world electricity by 2050," press release, International Energy Agency, May 11, 2010. References to the *IEA Technology Roadmaps — Solar Photovoltaic Energy and Concentrating Solar Power*.

27 "Atmospheric Brown Clouds: Regional Assessment Report with Focus on Asia," United Nations Environment Programme, 2008.

28 Source: Natural Resources Canada, which estimated the average number of kilowatt-hours per year per kilowatt of installed solar PV for major cities across Canada and worldwide.

29 NIMBYism, which stands for "not in my back yard," is rampant in the Internet age where misinformation can spread just as quickly as facts. This is best illustrated with protests against wind turbines and talk of a made-up condition called Wind Turbine Syndrome that has no basis in science.

30 Alexis Madrigal, "Hurricane-killing, space-based power plant," Wired.com. April 17, 2009. www.wired.com/wiredscience/2009/04/weathermod/.

31 National Space Society blog, June 8, 2010. http://blog.nss.org/?p=1821.

CHAPTER 3

A New Spin on Energy

Turning Waste Heat Into Tornado Power

*"The energy released by a large hurricane
can exceed the energy consumption of the
human race for a whole year, and even an
average tornado has a power similar to
that of a large power station. If only man-
kind could harness that energy, rather than
be at its mercy."*
— "The Power of Spin," *The Economist*, September 29, 2005

Tornados have always amazed me, but they've also been a source
of recurring nightmares. In my dreams I see them forming
outside the window of my house, or as I walk down a dirt road
in the countryside. I try to run, but the menacing things always
seem to follow me. Maybe the tornado scene from *The Wizard of
Oz* left a deep impression on me as a youngster. I can remember
being terrified as a 15-year-old by news coverage of the devastating tornado north of Toronto in Barrie. On that day in 1985, a

total of 13 tornadoes created mayhem across southern Ontario, two of them rating an F4 on the Fujita scale. This meant their wind speeds surpassed 330 kilometers an hour, and one of the tornadoes carved a path of damage more than 400 meters wide. Of the 12 who died, most received lethal blows after being lifted into the air and tossed back to the ground.[1] Dozens of people were injured by flying debris and hundreds of homes and businesses were torn apart. I could only imagine what it was like to live in Tornado Alley, America's most active tornado zone, an area between the Rocky and Appalachian mountains that stretches north from central Texas all the way to the Canadian prairies. The top 10 deadliest tornadoes in the United States took place between 1840 and 1954, killing about 2,500 people and injuring more than 9,000. Death tolls have dropped dramatically since the 1960s because of improved forecasting and early warning systems.[2] Still, even today's warning systems only give you 13 minutes advance notice, and 70 percent of the time they're false alarms.[3] Residents of Joplin, Missouri, had heard so many tornado warnings during a five-day period that they were caught off guard on May 22, 2011, by a ferocious twister that wiped out a third of their town. They had received a generous 20 minutes of advance warning, but it didn't seem to help. The kilometer-wide tornado, labeled the deadliest U.S. twister in more than six decades, killed at least 140 people.

When Tesla was in his late 70s, he wrote "Breaking Up Tornadoes," an article published in 1933. In it, he argued that it was possible to use explosives to disrupt the airflow of twisters and ultimately break them up. "It is entirely within our power to destroy them, or at least render them harmless," he wrote, suggesting that the government form an agency dedicated to tornado busting. "There is no doubt that, if such an undertaking were inaugurated, and many minds set to work, effective methods and means would be eventually developed and great loss of life and damage to property prevented."[4] The study of tornado control continues to this day. Scientists in Russia put out a study in 2009 that described the various methods that could be used to tame twisters, including heating up their tops — kind of what Solaren suggests it can do through microwave beaming of hurricanes — or cooling down their bases. They also pay homage to Tesla by mentioning his "effective" explosive method.[5]

There has been so much thought put into anticipating, avoiding, and destroying tornadoes that I became quite intrigued in early 2007 when I received the following email from a gentleman named Brian Monrad, a lawyer who was writing on behalf of a friend. "An engineer in Sarnia, Ontario, says he can create an artificial tornado and use it to produce electricity," he wrote. "Three top scientists from Oxford, Cambridge, and MIT agree that this is a real possibility and have joined the advisory board." Curiosity got the best of me, and a week later I found myself making the four-hour trek to Sarnia, where I was welcomed into the home of retired refinery engineer Louis Michaud, inventor of the patented atmospheric vortex engine concept.

GIVING IT A WHIRL

Michaud, a grandfather of four, is a slight and soft-spoken man in his late 60s who lives in a 1950s-style bungalow in a quiet area. Heavily industrialized Sarnia — nicknamed Chemical Valley because of the high density of chemical companies and refineries it hosts — neighbors a town called Petrolia, home to North America's first commercial oil well and touted by locals as the cradle of the global oil industry. Michaud worked more than 25 years there as a process control engineer at Imperial Oil, which is majority owned by ExxonMobil and ranks as the largest petroleum company in Canada. A spare-time inventor for most of his career, it was only after Michaud retired in 2006 that he could throw himself into the idea of extracting energy from tornadoes. It was then that he formed a company called AVEtec Energy with an aim to commercializing his unique vortex engine.

Minutes after we met, he took me to his garage to demonstrate a small-scale version of what he'd like to build, one day, on a massive scale. In his garage, between hanging bicycles and a tool bench, sat a hollow plywood cylinder about two feet high and nearly four feet in diameter. It had a round opening at the top, just big enough to stick your head into. At the bottom was a heating element connected to a small propane tank. Michaud fired up the propane heater and then lit a couple of pieces of saltpetre to get them smoking. As the heater began warming the bottom of the cylinder the air within it began to rise, carrying with it the smoke from the burning saltpetre. Within seconds, the rising saltpetre smoke started swirling and took on the shape of a vortex — a miniature tornado. A few minutes later, Michaud turned off the propane and opened the door to his garage to let out all the smoke that had accumulated inside. A woman walking her dog passed by on the sidewalk and looked on curiously; she probably thought we were smoking marijuana in the garage. I felt like Jeff Spicoli in *Fast Times at Ridgemont High*.

The demonstration, though interesting, could have easily been a high-school science fair experiment. Back inside the house, Michaud fixed me a cup of tea and began explaining his vision. The garage demonstration, he said, was just to show how heated air can be made to spiral as it rises. His idea is to build a structure that would support the creation and control of a tornado-like vortex that stretches many kilometers into the sky and spins at more than 300 kilometers per hour, as powerful as an F4 tornado capable of serious destruction. Why on Earth would anyone want to do such a thing? To generate clean electricity. A tornado needs to sweep in warm air from ground level in order to survive, so any supporting base structure would need air-intake ducts, and those ducts could be equipped with turbines. The powerful suction of air through those ducts would spin the turbines and generate electricity, Michaud explained. "I'm talking about a 200-megawatt device here," he said. "It would be 200 meters across and the vortex would be one to 20 kilometers high. It would have 10 turbines installed around its base, each producing 20 megawatts."

The key, he said, is to find a large source of low-grade heat that can get the tornado going, just like the propane heater he used in his garage prototype. One dependable source is waste heat from thermal power plants. Nuclear and coal power stations, as well as certain types of natural gas–fired plants, convert only one-third of their fuel into electricity. The rest of the energy in that fuel gets rejected as waste heat. That means a plant that produces 500 megawatts of electricity is throwing away the energy-equivalent of 1,000 megawatts of heat. That heat is released into the air using dry or wet cooling processes, or it can be absorbed into water and discharged into a large lake. It's a ridiculous situation, said Michaud. "We produce warm, humid air and just throw it away. In fact, we put fans on top of these cooling towers to get rid of the air, and we use a lot of power to drive the fans." He's not kidding: just running fans alone on a mechanical draft cooling tower eats up about three percent of a generating station's power output.[6] Michaud said that waste heat can be used to fuel a man-made tornado capable of generating 200 megawatts, which would improve the efficiency of a thermal power plant by 40 percent and offset the need for and cost of a cooling tower. Presumably, waste heat from solar thermal power plants could also be used. "The process could be developed with relatively little engineering effort," according to Michaud's business plan.

All of this might sound impossibly ambitious, but it's just the start. Longer term, Michaud sees the potential of building his vortex engines along tropical coastlines or on floating ocean platforms along the equator. Heat could be endlessly extracted from tropical waters as a way to sustain power-producing tornados. These same warm waters, which can reach as high as 32 degrees Celsius, are the source of energy for hurricanes and spontaneously appearing water spouts. "The passage of a hurricane can reduce surface temperature by up to 5 degrees Celsius," said Michaud. "They basically carry heat away from the surface."

Man-made tornadoes would achieve the same goal, drawing heat from the surface through the center of their vortex and dumping it into the upper atmosphere where it's more than 60 degrees Celsius cooler and the heat can more easily radiate into space, beyond the confines of the greenhouse effect. The only difference is that man-made tornadoes could generate electricity

at the same time. In a way, Michaud's vortex engines would operate like massive air conditioners for a warming planet but ones
that produce, rather than consume, electricity. "It's the first energy
source to contribute to global cooling, not global warming," touts
Michaud's business plan. Two months after meeting Michaud, I
wrote a story about his invention in the *Toronto Star*.[7] It captured
the attention of many readers, including Steven Levitt, professor
of economics at the University of Chicago and co-author of the
bestselling books *Freakonomics* and *SuperFreakonomics*. "This is
probably too good to be true, but all you need is one big idea like
this to work," he wrote in his *New York Times* blog. "Technology
and human ingenuity have solved just about every problem we've
faced so far; there is no obvious reason why global warming
shouldn't succumb as well."[8]

WHAT GOES UP . . .

The idea of harnessing heat from warm ocean waters has been
around for more than a century. As mentioned in the introduction, Nikola Tesla wrote in detail about extracting heat from
the ocean to generate electricity, a method of power generation
called ocean thermal energy conversion (OTEC) that Lockheed
Martin has pursued for the last four decades. Another idea that
has been around for several decades is the manipulation of atmospheric convection to produce electricity, which is essentially
what Michaud is attempting. Convection is one of three major
ways that energy is transferred between the Earth's surface and
the atmosphere. The other two are conduction and radiation.
Conduction happens when heat moves from a higher temperature substance to a lower temperature substance, for example,
from warm air into cold soil or warm soil into cold air. Radiation
is a transfer of heat through electromagnetic waves, which you
feel when the sunlight hits your face.

Convection, on the other hand, occurs when hot less-dense
air rises and cold dense air sinks. Convection is why a hot-air
balloon rises, and why macaroni in a pot of boiling water moves

up and down — the hot water closest to the heating element rises to the surface, cools, and then sinks back down in a constant churning motion. (The convection, or heat transfer, gets stronger as the temperature difference between the bottom of the pot and the surface of the water gets greater.) The lower atmosphere of our planet behaves, generally, in a similar way. The sun heats the surface, causing surface air to warm up and rise to much cooler layers of the atmosphere, where the air cools and descends back to the surface.

This is all somewhat orderly chaos until some sort of disturbance leads to storms, which can trigger the formation of tornadoes. With the right wind conditions, a rising pocket of warm humid air can begin to swirl, resulting in a spinning column that narrows as it starts drawing in warm air from the ground. As it narrows, it gathers speed and strength — the same way a figure skater can spin faster by pulling her arms tight against her body. The powerful tornado that eventually forms becomes a path of least resistance through which warm air on the ground can travel kilometers into the sky to the sub-zero temperatures of the upper troposphere. By sub-zero, I mean between –30 and –60 degrees Celsius. (Keep in mind, this is a simple explanation of an otherwise complex weather process that is still not fully understood. But the role that convection plays is clear.)

One of the first individuals to think up a way to capture the energy from this convective process, and certainly the first person to receive a patent on a design, was a Frenchman by the name of Edgard Nazare. Not much is known about Nazare, but it was between 1940 and 1960 that the French scientist, stationed at the time in Algeria, observed small sand tornadoes called dust devils in the south Saharan desert. Nazare, an aeronautical engineer and specialist in fluid dynamics, was captivated by the natural forces that caused these sand twisters to form. Convinced it was possible to replicate and control the process, he imagined a man-made apparatus called an "aerothermal power station" that could draw in warm ambient air from the desert floor and coax it to spin and accelerate as it ascended through a 300-meter-tall tower structure.

The tower narrowed at its center, forcing the air to start swirling — the same way water starts to spin as it goes down a drain. As with Michaud's vortex engine, Nazare's cyclone generator would produce electricity from turbines installed within the tower.

Nazare received a French patent for his invention in 1964, and while he championed the idea throughout his career, his efforts were met with resistance and ridicule. He became increasingly frustrated watching France aggressively pursue nuclear power while it ignored what he considered a much safer and morally superior technology. "Why do we systematically ignore the proposition to study aerothermal power stations? Every new technology has question marks," Nazare lamented in France's *L'Ère Nouvelle*, an obscure magazine that he regularly contributed to throughout the 1980s and '90s until his death in 1998.[9] As discouraged as he was, Nazare did prove to be an inspiration for others.

In the early 1980s, for example, the German government funded the construction of a "solar chimney" in Manzanares, a rural municipality about a one-hour drive from Spain's capital, Madrid. The $5-million pilot project didn't create tornadoes, but it did take advantage of convection as a way to generate electricity, in this case up to 60 kilowatts, or enough to power as much as 600 incandescent light bulbs. The design was simple, if not very practical. An area the size of eight NFL football fields was blanketed with a transparent sheet of plastic, just one-tenth of a millimeter thick. A grid of metal poles lifted the plastic from the ground, just high enough to walk underneath. At the center of this plastic-covered area was a circular metal chimney about 200 meters high and 10 meters in diameter. The chimney had a turbine, similar to a wind turbine, installed in the opening of its base.

During the day, sunlight would shine through the plastic sheet and heat up the ground, causing air trapped between the ground and the plastic to warm up by about 20 degrees Celsius. The plastic created a greenhouse effect. The warm, expanding air would then be channeled through the chimney, reaching speeds of up to 50 kilometers an hour. In essence, the chimney became a vertical wind tunnel. Air rushing straight up the chimney would spin an electricity-generating turbine installed at the base. The pilot project worked for about eight years until the chimney support wires rusted out and the chimney was blown over in a storm, but it did prove that the concept was scientifically sound. In fact, the solar chimney even produced electricity at night. The sun would heat the ground during the day, and the heat, stored in the soil, would continue to radiate from the ground in the evening, warming up the air under the plastic sheets and sustaining the upward flow of air through the chimney.

Overall, the solar chimney concept has much to offer. It is unique in its ability to convert low-grade heat into usable energy. A system can be built using cheap and abundant materials, such as plastic and metal. It is simple to operate and maintain. It can supply a certain amount of steady (baseload) power. It also doesn't require fancy solar cells, complex mirrors, or expensive water-cooling systems. But the approach has its problems as well. To get meaningful amounts of power out of such a system — that is, to get megawatts instead of kilowatts — requires construction of a substantially taller chimney. The taller the chimney, the stronger the updraft; the stronger the updraft, the more power you can generate. (It's similar, in a way, to water power. If you have two waterfalls of similar volume, the taller of the two has the potential to produce more power.)

In 2001, a company called EnviroMission announced that it planned to build a 200-megawatt solar chimney in southwest Australia that could generate 4,000 times more power than the Manzanares system. But to get that kind of power, EnviroMission must construct a solar chimney that's 130 meters in diameter and 1,000 *meters tall*, which is more than 11 times the height of the Statue of Liberty and nearly twice as tall as Toronto's CN Tower. Keep in mind that the world's tallest structure, the Burj Khalifa skyscraper in Dubai, is 828 meters tall, meaning EnviroMission would have to break a world record for its solar chimney to become a reality. Also, to create enough warm air to flow through that chimney will require a plastic- or glass-covered solar collection area of up to 35 square kilometers, roughly equivalent to

5,000 NFL football fields. The project is expected to cost nearly $1 billion.

EnviroMission has two more solar chimney projects planned for Arizona and, in 2011, secured $30 million in funding to cover early development costs. Solar chimneys are also being considered in other parts of the world. Projects are on the drawing board in the European Union and Africa, including a solar chimney in Namibia dubbed "Greentower" that would reach 1.5 kilometers into the sky and require a solar collecting greenhouse covering 37 square kilometers. None of these commercial-scale projects have reached construction phase, and there's no certainty they will.

ENTER THE VORTEX

Louis Michaud wasn't aware of the Manzanares project until a year or two after it was built, but he was quite familiar with the solar chimney concept. The idea had been floating around since the early 1900s, when Spanish Colonel Isidoro Cabanyes first described his "solar engine project" in a magazine article published in 1903.[10] In 1931, German author Walter de Haas, writing under the pseudonym Hanns Günther, proposed the use of solar chimneys — also known as solar updraft towers — as a way to generate electricity in a world that he believed would run out of coal.[11] Michaud was delighted to see the Manzanares project built, because it demonstrated how warm, expanding air could be converted into mechanical energy as it is carried upward by convection. But in his mind, solar chimneys are inefficient and not very practical, specifically because of the dramatic heights needed to generate megawatts of power. "The cost of the chimney is just incredible," he once told me.

This is part of the reason Michaud, some time in the mid-1970s and while working in a Quebec paper mill, turned to the idea of creating tornadoes to generate electricity. It was a hobby, really, a personal project that grew out of his side interest in meteorology and his on-the-job exposure to convective processes in an industrial setting. He knew that tremendous amounts of energy could be harnessed through atmospheric convection and saw the solar chimney as one way of doing it. At one point, he even considered the use of a chimney-like tube held high in the sky by a blimp, but it just wasn't practical or efficient enough. Then it struck him: tornadoes are their own chimneys. Not only that, these self-swirling columns of air can stretch several kilometers high.

If you think about it, the purpose of a physical chimney is to prevent cooler, higher-pressure ambient air on the outside from mixing with lower-pressure warm air as it flows up through it. The high-pressure air desperately wants to mix with the low-pressure air, but if this happens it will weaken or kill the convection process. The beautiful thing about a vortex is that it doesn't need this physical separation. Instead, the centrifugal force created from its rapid spinning makes an air wall that effectively pushes back on the surrounding ambient air that's trying to muscle its way inside. As a result, the column of warm air within the vortex can rise to great heights with little interruption. Michaud, realizing that the sky was literally the limit, turned his attention to making tornadoes.

It was soon after this realization that Michaud discovered the work of Edgard Nazare. Michaud was intrigued by Nazare's ideas and began corresponding with the French scientist in 1973 and through much of that decade. Back then, of course, they didn't have the advantage of the Internet and email, so their exchanges were in no way frequent. In fact, there was little contact between the two inventors during much of 1980s. Michaud was hired as a process control engineer at Imperial Oil (known under the trade name ESSO) and immersed himself in his new refinery job, to the point that he stopped publishing research papers and put aside his work on "tornado power." It still occupied his mind, however, and the new job proved crucial to refining his ideas. "One of the reasons I went to work at ESSO is that the petroleum industry had a lot of applicable techniques for what I was studying," Michaud recalled during an interview in spring 2010. He said a large petroleum refinery might have half a dozen cracking furnaces, 10 compressors, and 6 distillation towers. It was the perfect learning ground. "You're able to use those assets, and surprisingly there are a lot of convection processes there. You end up learning from the real world, rather than just computer simulations."

Fast forward to 1992. Michaud checked the mail one day and found a large brown envelope sent to him by Nazare. "It was all his papers," said Michaud, who was in his late 40s at the time and didn't realize that Nazare was well into his 70s. "He was passing on the torch to me, I guess. He figured I was the person most likely to bring the concept forward." Michaud believed that Nazare, having worked on the concept for most of his life, was disappointed that it never gained mainstream acceptance. "He was looked at as a bit of an odd ball. He had to self-publish and after a while got frustrated. It was good for me to see what happened to him, because I realized I had to make sure I explained the rationale behind what I was proposing." Michaud said one tool Nazare never had during his career was the Internet, which would have made his work more accessible and given him a way to collaborate with like-minded individuals around the world. Without that community, Nazare remained a scientific renegade within France's conservative scientific establishment.

Michaud felt a personal obligation to Nazare to carry the concept forward. He continued to refine his ideas through the 1990s, but it was only during the lead-up to his retirement in the first few years of this century that he began to promote the commercial potential of tornado power. He drew up a detailed design for a utility-scale vortex engine in 2001 and filed for a patent the same year.[12] In 2005, a year before his retirement, he created a website, VortexEngine.ca, and populated it with as much explanatory information as possible as part of a larger effort to raise awareness of the idea. He developed a business plan and started work on a small-scale prototype. Michaud also knew he needed some credible backing, so he reached out to the world's top climate and atmospheric scientists, many of whom agreed to be informal advisors. One of those scientists is leading hurricane expert Kerry Emanuel, a professor of meteorology at the Massachusetts Institute of Technology. Emanuel, who has read Michaud's papers and is in occasional email contact with the Sarnia engineer, said in an interview that there's no doubt in his mind the vortex engine will work. "It's mechanically very simple, and that's the beauty of it," he said. "One of the nice things is that it achieves a thermodynamic efficiency you wouldn't expect, with the advantage that it can operate on low-temperature waste heat." The proposed Australian solar chimney, for example, can convert at most about 3 percent of the thermal energy it collects into electrical energy, while Michaud's vortex engine can convert closer to 20 percent. "So, for sure, this is something worth exploring," said Emanuel.

FORGING ON

By the time I met Michaud in spring 2007, he was working full time on his vortex engine and trying to raise money to construct prototypes that were much larger than the small demo in his garage. He had managed, through an Ontario government program, to partner up with the University of Western Ontario to run computer simulations, but that wouldn't be enough to excite investors. People want to see real-life tornadoes, not computer screens. But taking his idea from a one-meter-diameter garage prototype to a commercial power facility up to 200 meters across? Building a vortex engine capable of producing a tornado over a kilometer high? Well, that wouldn't so be easy. For one, it requires a significant injection of capital. Just building a four-meter-diameter prototype would likely cost $300,000. A pre-commercial model between 10 and 50 meters in diameter, connected to a large steam source and equipped with small power-generating turbines, might cost several millions of dollars. At this size, Michaud has wondered whether an amusement park, such as Walt Disney World, might want to build one of his tornado machines as a tourist attraction.

Any larger and it gets *really* expensive, though still a deal compared to a solar chimney. Michaud has pegged the price of a full-scale facility about 100 meters in diameter at roughly $60 million. At that size, it would generate 200 megawatts of power using waste heat ejected from a thermal power plant fueled by coal, natural gas, or uranium.[13] To build an even larger standalone facility that could extract heat from tropical ocean waters would cost north of $200 million. This tropical facility would be capable of generating up to 500 megawatts and would have a diameter of 200 meters, allowing for the creation of a monster F4 or F5 tornado.

It sounds like a lot of money and it sounds a little — well, maybe a lot — crazy. But what should matter more is the payoff. If you believe Michaud's calculations, his atmospheric vortex engine is an economic slam dunk. He figures it could generate electricity for less than 6 cents per kilowatt-hour, and possibly as low as 3 cents per kilowatt-hour, making it a better deal than nuclear and natural gas, far cheaper than renewables such as wind and solar, and competitive with the dirtiest of coal plants. The capital cost of building his vortex engine facility would be about $300,000 per megawatt, or about a quarter of what it would cost to build a new coal plant of the same power output. More than that, it eliminates the need for a conventional cooling tower that, at a new power plant, would cost just as much, if not more, to build. "From an engineering perspective, building this is not really a big deal," said Michaud. He firmly believes his power-generating vortex engine is one of the lowest-cost options for producing electricity, particularly zero-emission electricity.

Since our first meeting in 2007, I've checked in with Michaud occasionally for progress reports. He's still pushing, still networking, still spreading the word, but not much has happened in four years. The largest prototype he has built to date is only four meters in diameter; it was tested in summer 2008 and created twisters more than 12 meters high. Decent enough, but like any new technology that requires an openness of mind and a willingness to take risk, finding the public granters or private investors who can take the project to the next level has proved to be a major hurdle, at least for now. Of course, there are some small exceptions. One was a lady in British Columbia who read about Michaud's tornado power concept in the San Francisco-based magazine *Ode*. She took a $100 "carbon tax" rebate check issued by the B.C. government and decided to sign it over to Michaud. "I think we should be putting as much money as possible into alternative energy research, so I would like to give my $100 to you for your project," she wrote him.

The media, of course, love Michaud. Who can resist writing about a retiree trying to make powerful tornadoes to supply electricity to towns and cities? Unfortunately for him, coverage hasn't translated into financial support or a willingness on the part of any utility to give his vortex engine a spin. "It's not an incremental step from somebody saying, 'This is interesting' to somebody saying, 'I'm going to invest in this,'" said Michaud. In other words, the vortex engine is fascinating enough to open doors, but too "out there" to get signed commitments. "The utility industry is very conservative. I was contacted by the Tennessee Valley Authority after one article, and the manager of technical innovations there was quite interested. I prepared a presentation and sent it, but once she talked to her boss it was the end of the discussion." Michaud opened up a similar conversation with Alberta power utility TransAlta, which operates a 500-megawatt natural gas plant in Michaud's hometown of Sarnia. It ended the same way after the local business manager escalated the idea to corporate headquarters in Calgary. "The CEO has got to get behind it if it's going to go anywhere," said Michaud. "But once you get past the engineers, the executives aren't interested."

Scientists, on the other hand, seem to gravitate toward the concept. A sign of the times, maybe? Nazare, you'll recall, was shunned in his day by France's scientific community, a time when climate change, peak oil, and energy security in general weren't pressing mainstream concerns. They are now. Michaud had no problem attracting the attention of the U.S. Department of Energy's Sandia National Laboratories. He traveled to New Mexico in January 2009 to present his idea to a room of 10 Sandia scientists eager to learn more. "They were awed," recalled

Michaud. "One of the directors there, Sheldon Tieszen, told me directly he wanted to personally work on this thing." Such a partnership with a respected government lab would have been invaluable to Michaud, and indeed, Sandia applied for funding through ARPA-E — the U.S. government's cash-flush Advanced Research Projects Agency (Energy) — so it could demonstrate Michaud's cyclonic creation as part of a CO_2 capture system that could generate electricity at the same time. Their application was unsuccessful.

REALITY CHECK

I think it's time to tackle the elephant in the room, and it's a big elephant, right? Here we have a man who wants to create tornadoes in order to generate power. But tornadoes have killed thousands of people in North America over the past hundred years and caused billions of dollars in damage. They are powerful, destructive, unforgiving, unpredictable, and feared. So the biggest question isn't whether we can create tornadoes, it's whether we can *control* them. And even if Michaud makes a convincing case that they can be controlled, and that the risks are extremely slim of a tornado hopping its concrete pen and terrorizing a town like Godzilla let loose on Tokyo, does it even matter? Can the fear factor be overcome? Can the barrier of negative public perception be surmounted?

Seeking advice on how to move his invention forward, Michaud traveled to Toronto in spring 2010 to meet with Tom Rand, who heads up the clean technology practice at the MaRS Discovery District, a kind of public-private incubator for technology ventures. Rand acts as an advisor at MaRS; in his private life, he's a big investor in and champion of clean energy and green technologies. He's also a practical man, someone who has run his own high-tech businesses in the past, so when presented with Michaud's idea, he was intrigued but doubtful of its commercial viability.

"I'm sure in theory it works," he told me shortly after meeting with Michaud. "But seriously, we're going to control cyclones that reach into the sky? People have a hard time visualizing the risks that might be involved and the regulatory hurdles. The endgame is hard to imagine. Building hundreds, even thousands of these things is hard to imagine. You have to imagine thousands of these man-made cyclones being controlled. But how do they affect the upper atmosphere, air routes, and weather systems, among other things? It may not be a big deal in the end, but when you look at this you're thinking, 'Holy crap, that's a lot to get through. I'd rather just drill holes and take energy out of rock.'"

That reticence Rand described translates into that innovation killer called risk. Michaud can invent a better, cleaner, and cheaper way to generate electricity, but what matters most is whether big utilities and big banks can accept the risks, said Rand, even if those risks aren't as high as they seem. Banks, he said, are "allergic" to technology risk. Utilities, Michaud has already learned, are culturally resistant to change. Inventiveness and risk-taking are not part of their DNA. As Lynne Kiesling, an economist at Northwestern University once said, the "electric power industry is one of the least innovative industries in the modern economy."[14] Proven reliability is what matters most, and efficiency — even cost — is routinely sacrificed to assure that new equipment or processes or systems will not fail. "There's a psychological momentum out there that things, as they are, are largely working in the world," said Rand. "So there's no reason for anyone to put their career on the line to support something new." It's difficult enough, he added, building support for less risky, better-proven options already on the table, such as geothermal, concentrated solar thermal, and offshore wind. "Inventors can think they're solving a problem, but they're not solving a problem. They're just putting another option on the table. There are already lots of people who can make clean power and at less risk, and their products are not getting installed. So I often ask the people I advise, what makes you think what you have can get installed?"

It's not difficult to get Rand worked up about this issue. He, like many of the inventors and entrepreneurs he advises, is frustrated with the chasm that exists between pure innovation on one side and deployment of that innovation on the other. "The people who get paid to make our machinery run, they get paid to keep it running, not to change direction. Who's going to stick their necks out? Why would you stick your neck out? Imagining hundreds of cyclones creating energy is not going to help unblock those institutional blockages. These days you need to de-risk, de-risk, and de-risk some more to make it all easier to swallow. Michaud's invention, this just does the opposite, and sadly, it may just be an issue of perception," he said. The bottom line on this project? "This will scare the shit out of bankers."

Michaud understands the concerns, and while he finds them exaggerated, he also understands that perception, the same one that kept Nazare from realizing his dream, is powerful. Still, many of our activities on this planet are dangerous and highly risky, be it sending workers kilometers underground (the 2010 Chilean miner saga) or going deep drilling offshore for oil (BP's Gulf of Mexico spill) or building containment ponds for toxic sludge (Hungary's "red mud" flood of arsenic, lead, cadmium, and chromium from an aluminum ore processing facility). Capturing carbon dioxide and storing it underground is untested on a large scale and considered highly risky given the consequences of a major leak, yet major utilities have embraced this as a way to mitigate climate change. Michaud has worked most of his career in refineries, which are constantly at risk of major explosions and toxic gas leaks. Described by some as "barely controlled bombs," oil refineries need multiple backup systems and 24-hour supervision to keep them from setting off. Yet we live with them. As a society, we've also become quite tolerant of nuclear fission power despite the highly radioactive waste it creates and the ever-present risk of a radiation leak or Chernobyl-type disaster. Japan's Fukushima crisis might be a setback for nuclear, but there will remain a resigned acceptance of it.

"At first they think my vortex engine is crazy. Then once they believe it's possible, they start saying it's dangerous," said Michaud. "But I don't think it's dangerous at all. It's actually relatively simple technology, and compared to nuclear power, it is peanuts." The reason Michaud believes his vortex engine is safe is because its lifeline — that is, the heat source that starts it up and keeps it alive — can be easily controlled or severed. A typical vortex engine facility, one large enough to produce 200 megawatts, would be contained within a round concrete or steel arena more than 100 meters in diameter and about 50 meters high — no larger than a typical natural draft cooling tower used already by most of the world's nuclear plants. The tornado created would rise out of the center of the arena and reach more than a kilometer into the sky, but it would get its energy through 20 or so air-intake ducts equally spread around the arena's base. Upon startup, fans would be used to blow warm, humid air from the waste heat into the bottom of the arena at an angle, so that the warm air starts spinning inside. As the spinning air rises, it begins reaching for the cooler ambient temperatures above, gathering momentum until a tornado forms. At that point, there is no longer a need to blow air into the arena; the tornado takes on a life of its own and begins sucking air into the arena through the intake ducts. As a result, the electricity-consuming fans that first blew air into the arena start operating like turbines that spin as the tornado draws air through them. As the turbines spin, they generate electricity.

Michaud said the air-intake ducts are the key to controlling the tornado: "The station can only let in so much air because the [air-intake] openings are only a certain size. Each opening has a damper that can be shut down." In other words, by restricting air flow into the arena, a utility could easily weaken or completely shut down its power-producing vortex engine at any time. I put the question to meteorologist Emanuel, who would know better than anyone else the risks involved with Michaud's vortex engine. "Only under very extraordinary situations would it be a problem, and those are the circumstances in which a normal tornado might form anyway," Emanuel told me.

COMFORT FACTOR

Emanuel did have one concern: "I might worry a bit about how a rotating updraft might pose a risk to aircraft. I wouldn't want to fly a hot-air balloon into that." But even here, the risk may be negligible. More than 1,000 tornadoes form each year in the United States, according to the National Oceanic and Atmospheric Administration, and predicting when and where they will form is virtually impossible. This unpredictability would seem a greater danger to aircraft than man-made tornadoes that can be easily controlled and clearly mapped out for all aviators to see.

There are many other possible objections, ranging from concerns over visual and noise pollution to the impact on birds and bats. Michaud said the amount of debris caught up in a tornado determines to a large degree its loudness or visibility. It's generally understood that much of the destructiveness of a tornado has to do with the dust and debris that whips around within its powerful vortex. Leaves, dirt, broken tree limbs, hail, fragments from destroyed buildings — even farm animals and vehicles — all of this material captured by a twister would explain its menacing look, as well as why it might sound like a freight train coming at you from a distance. The only thing visible in Michaud's vortex engine would be some water vapor, making it a bore by comparison. "You might actually want to have a lightshow on it at night with different colored lights to make it stand out," said Michaud. As for noise, he said much of it would be contained within the arena facility. "You could also find other ways to muffle it."

Birds and bats? Guesswork wouldn't be an option. Understanding the impact there would require detailed study of full-scale prototypes, but to do it means those full-scale prototypes need to get built, creating a classic chicken-and-egg standoff. There's also the weather modification element of Michaud's invention. It can, after all, encourage precipitation to fall by creating moisture-filled funnel clouds above it. If you live in drought-inflicted regions such as the U.S. Southwest or Australia this might be appreciated, but not so much in England or other places that suffer from too much rainfall. Many people would consider this kind of "geoengineering" forbidden territory, on par philosophically with cloning humans and or creating new life forms in Petri dishes.

But beyond such controversies, there is an overarching question here. Is the opportunity too big to ignore? Are the risks, perceived or otherwise, worth the benefits that Michaud's vortex engine has the potential to bring? Consider that there are nearly 1,000 thermal power plants in the United States, mainly fueled by coal, natural gas, and uranium, in that order. They collectively supply about 3,500 terawatt-hours of electricity each year to the country, or about 88 percent of America's electricity. In Canada, which has large hydroelectric resources, thermal power contributes much less — about 41 percent of power supply or 250 terawatt-hours. Taken together, North America's thermal power plants generate about 3,750 terawatt-hours of electricity annually and waste just as much energy (or more) as heat, which is dumped into the air or lakes with potential to disturb local ecosystems.[15] Theoretically, vortex engines located beside these power plants could produce around 800 terawatt-hours, enough electricity to power Canada, New York, and Massachusetts combined. It would also mean 500 man-made tornadoes scattered throughout the continent and spinning 24-hours a day. If Michaud's vortex engine could be deployed in the tropics and proven to operate just on the heat in ocean water instead of waste heat from a power plant, then the future possibilities — while difficult to imagine — are endless.

Is it just a matter of becoming comfortable with the technology? Michaud hopes that's the case. If he had $1 million (he doesn't), he figures he could build a sizeable vortex engine about 15 meters in diameter that could create a small tornado capable of reaching hundreds of meters into the sky. It might even produce a small amount of power, but its main purpose would be to show people what it would look like, how it would behave, and how it could be controlled. It might make the public more accepting of the technology, but that's just one of many big steps required to take tornado power mainstream. "There will always be deniers, like those who thought we would never need more than five computers in the world," said Michaud. But his bigger concern is with those who may actually like the idea but aren't motivated or brave enough to get behind it. "Utilities just don't have that mentality to pursue this. It forces you to ask: who is there to work with?"

Government could play a larger role, but it isn't. If Michaud was personally wealthy, he could self-finance construction of a full-scale commercial prototype and test it on some isolated island in the South Pacific, but this isn't an option for a man surviving on a pension. An angel investor, a visionary philanthropist, could step in with the millions of dollars needed to take the vortex engine to the next level. Odds are, however, that the vortex engine will remain a curiosity that, like so many inventions of the past, never gets the chance to prove its worth. Whatever happens, Michaud is determined to move forward. "I could work as a consultant and get more money for my effort, but this is something I like doing," he explained. "I figure, if you realize there's a potential there and nobody else is doing anything about it, it's just not right for me to say, 'Okay, nobody is listening to me, I give up.' I'm prepared to keep going."

Some may feel it's a bad idea for Michaud to mess with Mother Nature. Others will argue that Mother Nature has a lot to teach us about solving human-caused problems. Jay Harman, as you'll read in the next chapter, would count himself among the "others." He has devoted his life to observing and emulating nature, and after years as a lone runner, his efforts are beginning to pay off.

Notes:

1 Anne O Carter et al., "Epidemiologic Study of Deaths and Injuries Due to Tornadoes," *American Journal of Epidemiology* 130, no. 6 (1989): 1209–1218.

2 U.S. Department of Commerce National Climatic Data Center (data updated to April 2008).

3 Samantha Davies, "Largest Tornado Study in History Taking Place in Tornado Alley," NBC Dallas Fort Worth affliate website. May 13, 2009. http://www.nbcdfw.com.

4 Nikola Tesla, "Breaking Up Tornadoes," *Everyday Science and Mechanics*. December 1933.

5 A. Yu Varaksin et al. "Controlling the Behavior of Air Tornados," *High Temperature* 47, no. 6 (2009): 836–842. (Authors are from the Joint Institute for High Temperature at the Russian Academy of Sciences.)

6 "Site Preparation, Construction and Operation of the Darlington B Nuclear Generating Station," environmental assessment, Ontario Power Generation. April 12, 2007. 13.

7 Tyler Hamilton, "Taming Tornadoes to Power Cities," *Toronto Star*. July 21, 2007.

8 Steven D. Levitt, "Are Man-Made Tornadoes the Answer to Global Warming?" Freakonomics.Blogs.NYTimes.com. August 13, 2007.

9 Edgard Nazare, "L'Homme Peut Faire Des Cyclones et Dompter Leur Énergie," *L'Ère Nouvelle*, no. 52 (1985). *L'Ère Nouvelle* translates to "The New Era." Created in 1980, the magazine was a forum to express alternative views that sought protection for the environment, individual freedom, human health, and dignity; Edgard Nazare contributed frequently until his death in 1998.

10 Isidoro Cabanyes, "Proyecto De Motor Solar," *La Energia Flèctrica* 8, no. 4 (August 25, 1903).

11 Hanns Günther's 1931 book was titled *In hundert Jahren — Die künftige Energieversorgung der Welt*, which translates to "In a Hundred Years: The Energy Future of the World."

12 Michaud now has a Canadian patent (CPO #2460564), an American patent (US PTO #20040112055), and a European patent (PCT Europe #W003025395) on his atmospheric vortex engine.

13 Some natural gas plants are "combined cycle" plants, meaning that they run a gas turbine and a second steam turbine on the high-temperature waste heat from the gas turbine. This can bring the efficiency of a natural gas power plant to about 45 percent, i.e., a 500 megawatt single cycle plant can be turned into a 700 combined cycle plant using the same amount of natural gas. Still, even this more efficient plant will waste about 800 megawatts-equivalent of waste heat. Michaud's vortex engine can harness about 160 megawatts from that 700 megawatt combined cycle plant, compared to 200 megawatt from a 500 megawatt single cycle plant.

14 Ronald Bailey, "Cogeneration: Producing Heat and Light and Profits," *Chief Executive Magazine*. January/February 2010.

15 China, by comparison, is expected to generate 3,700 terawatt-hours of electricity annually by the end of 2010, about 78 percent of it coming from coal.

CHAPTER 4

Copying Nature's Playbook

Capturing Efficiency Through Biomimicry

"Human subtlety . . . will never devise an invention more beautiful, more simple, or more direct than does nature, because in her inventions nothing is lacking, and nothing is superfluous."
— Leonardo da Vinci

Jay Harman may be the president of a company well positioned to disrupt the industrial world, but he's not your typical suit. That is to say, he doesn't like dressing up in business attire. He'd rather be hiking in the Western Australian bush or, even better, spearing pink snapper off some stunning coral reef, as far away as possible from corporate boardrooms and all the hobnobbing of industry conferences. One might say it's the boy inside the man. Harman spent most of his childhood during the early 1960s literally immersed in nature. He routinely ditched school to hang out on the beach, to snorkel, explore, and observe. At age 10, he was already an enthusiastic spear fisherman, and by 18 he was state junior champion. "When you're diving, there's no trace of humanity," Harman, now in his 60s, told me during an afternoon chat. "It's just you and the wild flying through a water-thick atmosphere." As a young adult, Harman worked for Western Australia's fisheries and wildlife department on coastal patrol boats and even became captain of his own research vessel. The ocean and the strange, fascinating animals and organisms within it became his second family. His second nature.

More influential on his work, Harman's intimate relationship with the ocean taught him as much about fluid dynamics and energy efficiency as any university education could. "Nothing is left still in the ocean," he said. "The medium is moving. The fish, animals, everything is moving. By watching this, I discovered there are commonalities in all motion in our universe. What I was able to see as a kid was the essence of motion." For example, he observed that sea kelp, a type of large seaweed, had this amazing ability to stay in one spot, seemingly fixed to the coral below it, despite being battered by the ocean surf during storms. At first he figured these long strands of kelp had strong roots that anchored them firmly to the coral, but one day he grabbed a strand and realized the seaweed could be pulled out almost effortlessly. He eventually realized that the ribbon-like kelp could hold its ground by changing itself into a spiral shape, which let the water rush past it with minimal resistance.

This spiral pattern stuck in Harman's mind, and he began noticing it more and more — both under and above the water. Look at the cross section of a nautilus seashell and you'll see a spiral structure. The cochlea of the human ear is a spiral shape. Sunflowers, pinecones, pineapples, and artichokes all have spiral patterns in their biology. From something as tiny as a DNA strand to water whirling down a sink drain to tornadoes and hurricanes to entire galaxies, the spiral shape is a recurring pattern in nature. Even the microscopic pores in our skin are spiral shaped. It occurred to Harman that this pattern is there for many reasons, one of which is to streamline the flow of anything fluid, including gases such as air and apparently even human sweat. Over millions, even billions, of years life has evolved to create ideal shapes that reduce their friction on the surrounding environment or take advantage of natural air and water flows. This is why Louis Michaud's vortex engine is so appealing; the air flow that both forms and fuels a tornado is a time-tested demonstration of nature's efficiency.

You may have heard of the Fibonacci sequence, named after the 13th century Italian mathematician Leonardo of Pisa, also known as Leonardo Fibonacci. The sequence begins 0, 1, 1, 2, 3, 5, 8, 13, 21, 34, 55 . . . — can you spot the pattern? Any number in the sequence is the sum of the two previous numbers. Expressed geometrically, the Fibonacci sequence forms a spiral, observed repeatedly in nature. Harman, while working at the department of fisheries and wildlife, studied these geometries and their relationship with natural flow. Realizing that spirals, not straight lines, were the path of least resistance in nature, he asked himself: can industrial equipment be designed to take advantage of these naturally efficient flow geometries?

During much of the 1980s and '90s, he didn't just try to answer that question, he quit his job at the fisheries department and set about designing products to prove that, yes, shapes found in nature have much to teach us about industrial design. This was preceded by a brief detour to college, where the Jesuit-raised 29-year-old decided to enrol in religious studies with a focus on eastern mysticism. It may seem like an odd path, but Harman knew that many religious symbols and icons throughout the ages have been based on naturally occurring shapes and patterns, including spirals. "In my own way, I was able to link this ancient recognition and knowledge of these patterns and what I'd observed in nature," he said. "It gave me a better understanding of how things function, and how to get the best out of technology." He traveled the world studying under a number of different religious teachers. Eventually, Harman ran out of money and at 32 returned to Australia, where he decided to start up a technology research firm with a $40,000 loan from a friend.

He gained valuable experience running a business, and two years later he successfully took the company public and sold his shares. Around the same time, Harman began designing high-performance surfboards, canoes, and boats based on the shapes of whales and dolphins. One design, his transparent Goggleboat, was the world's first all-plastic, seamless marine craft, while a series of rescue boats dubbed WildThing won awards for their light weight and durability.

Harman, like the ocean he had observed all his life, remained restless and relentless. Despite having great success with his boats, he was determined to improve their performance even further and, in 1991, set about designing a new type of propeller based on the spiral shape of water as it goes down a drain.

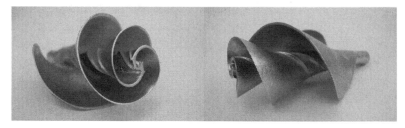

PAX's Lily Impeller, which truly mimics the shape of the flower it's named after.

Somehow he managed to freeze the whirlpool shape in a bathtub as the water drained, and he took a cast of that shape as a model for his prototype propeller. He called it the Lily Impeller because of its close resemblance to the X-ray image of a calla lily, which has a spiral shape. It blew away the competition. "Every field of movement in our universe shares the same algorithm that the whirlpool in the bathtub does," explained Harman. "Once I had designed that propeller, I then had fans, I had pumps, and on and on from there." Then living in California, he experimented for several years until 1997, when he and his wife, Francesca Bertone, founded the company PAX Scientific in San Rafael. Its corporate mission, according to its website, is to "bring the exceptional efficiencies of natural flow to fluid-handling technology" — everything from fans, propellers, and turbines to pumps, mixers, and heat exchangers. In other words, making anything that moves or is moved by fluids or gases do more work with less energy.

The idea of having a better, more energy-efficient design for something as rudimentary as a fan or propeller might sound, well, boring. In the world of energy technology, it certainly doesn't have the same headline-grabbing appeal as building solar-power plants in outer space. But the movement of air alone, as Harman is quick to point out, is an enormous enterprise: companies directly involved with airflow and air conditioning are nine times larger than the movie business and 16 times bigger than the recording industry. Given how many of these products keep the plumbing and engine of our global economy flowing and humming, the potential impact is astonishing. "When Jay first showed me what he had, I broke down about 20 sectors it would revolutionize," said efficiency guru Amory Lovins, co-founder and chairman of the energy think-tank Rocky Mountain Institute. Lovins was so impressed, he agreed to sit on a PAX advisory board. "What he has here could save over a tenth of the world's energy."

With that kind of efficiency claim, you'd think the world's industrial giants would be lining up to license the technology or place orders. Harman said interest has been slowly building over the past few years, but the idea of altering industrial systems to accommodate new energy-saving products modeled after the shape of a seashell initially creates more pushback than pull — much of it routed in fear of change and hidden behind a veil of skepticism. As Harman said during one conference he attended in 2008, "When you come along with a 'game-changing' design, you come across as a nutcase."[1] You come across as mad like Tesla.

MOTHER KNOWS BEST

Harman isn't the first or the only person to have designed and engineered products based on observations of nature. A famous example is Leonardo da Vinci, who used to sketch designs of hang gliders and other flying machines in the 15th century based on his own study of birds. "When da Vinci looked at a bird's wing, then attempted to design a wing that could carry a man aloft, he was practising a science that today we call bionics," wrote Philip Callahan in his 1975 book *Tuning In To Nature.* "In bionics, man carefully examines how nature works, and then tries to improve his own technology by copying nature. Da Vinci was one of the earliest and certainly the greatest practitioner of this science."[2]

This term bionics was coined by an American doctor named Jack Steele in 1958 and became a familiar word for anyone who watched the hit television show *The Six Million Dollar Man* during the 1970s. The main character, Steve Austin, is a former astronaut who is so badly injured in a crash that his legs, right arm, and left eye are replaced with bionic implants designed to look and operate just like real body parts, only stronger and faster. The show gave us the bionic man, bionic woman, bionic boy, bionic dog, even a bionic Bigfoot. Alas, it also gave the general public a narrow view of what bionic science was all about, that bionics was strictly about building life-like but superior body parts out of electronics and synthetic materials. That's why if you ask the average person about a commonplace product like Velcro, they're not likely to associate it with bionic science, even though that's exactly where the idea came from.

The discovery that led to Velcro demonstrates how a simple observation of nature can lead to a ubiquitous product we now take for granted. Swiss engineer George de Mestral took his dog for a hike in the summer of 1948 and came back home covered in seed-packed burrs, which were stubbornly clinging to his clothes and to the dog's fur. He put one of these burrs under a microscope and noticed that each of the prickles on it had teeny hooks on the end. Not only were these burr prickles capable of piercing de Mestral's clothing and his dog's fur coat, the hooks on the prickles allowed the burrs to hitch a ride as far as their unknowing carriers would take them.

De Mestral realized this was nature's intentional design. The plants flourished if able to disperse their seeds across a large area, so evolution favored those with burrs that could attach to and travel with passing animals. Mother Nature's solution to seed dispersion inspired de Mestral to create Velcro-brand fasteners, the solution for schoolchildren everywhere who don't yet know how to tie their own shoelaces (among other applications . . .). The name Velcro, by the way, is a hybrid of the words velour and crochet.

Bionics is one of many terms that have emerged over the years to capture the idea that imitating nature — its creatures, systems, processes, and forms — can inspire innovative answers to age-old problems or simply improve the way we do things. American biophysicist Otto Herbert Schmitt preferred the term biomimetics, which made it into dictionaries in 1974 but was introduced by Schmitt sometime during the late 1960s. Bio-inspired engineering is another frequently used but less popular term. In the late 1990s, however, the word biomimicry began to capture imaginations. The colorful science writer Janine Benyus came out with *Biomimicry: Innovation Inspired by Nature*, a bestselling book in which she described biomimicry as the "conscious emulation of life's genius" in a society too accustomed to taming or trying to improve nature, or even worse, extracting from it with brute force.[3] Mother Nature, Benyus argued, has supplied us with 3.8 billion years of research and development through natural evolution. Why wouldn't humanity borrow the well-seasoned recipes from her cookbook?

Biomimicry and bionics are not interchangeable terms; there is at least one important difference. Biomimicry, the way Benyus sees it, has sustainability as a key goal, meaning our purpose for imitating nature is to not just innovate for the sake of making cool products, but to come up with sustainable innovations that are respectful of nature. "Biomimicry uses an ecological standard to judge the 'rightness' of our innovations," she wrote. Also, innovation is more than just a cleaner way to generate electricity, fight disease, or make green materials and chemistry. Nature-inspired innovation can affect urban planning, government policy, and even political systems.

Green chemistry is one area where biomimicry is thriving, according to Benyus, who founded an organization called the Biomimicry Institute in 2005, which acts as a research clearing-house and promotes the study of biomimicry. I had a chance to sit down with the Montana resident during one of her visits to Toronto, where her institute helped set up a two-year course at the esteemed Ontario College of Arts & Design that links design and engineering with biology.[4] "The industrial recipe these days is all about using every element in the periodic table and brute force to turn them into what you want," she said. "It's all about high pressure, high heat, and high toxicity." Bio-inspired green chemistry, on the other hand, is about achieving the same objective but using nature's chemistry book. "The recipes are very elegant. They use fewer elements. Water is the solvent. And it's low heat." The bottom line: "Life does chemistry really, really well."

One example she gave is an Australian-based venture called Biosignal, which was founded in 1999 by University of New South Wales biology professors Peter Steinberg and Staffan Kjelleberg. They observed that a type of red algae found in Sydney's Botany Bay was unique in that it was never covered in a scummy layer of bacterial slime called biofilm. Biofilm forms when bacteria get organized: the first ones to land on a surface will send out chemical signals that attract others, kind of like how a spontaneous protest in the streets can start with just a few people sending out messages on Twitter. "When enough of them are there, they turn on their toxins and start forming a sugary biolayer called biofilm," explained Benyus, adding that the biofilm is what help keep the bacteria hidden from antibiotics, making it more difficult to fight serious infectious diseases such as cholera. So how do you keep a bunch of protesters from using Twitter to congregate in front of city hall? If you're an oppressive regime, you jam their smart phones. That's what the red algae does, Steinberg and Kjelleberg discovered. "The algae release a molecule, a furanone, into the seawater that is like bacteria's signaling molecule, but different in that it jams the receptors of the bacteria in the area. It jams the communications network of the bacteria," Benyus continued.

"So Biosignal has mimicked this, and they now have a cleaning product that can create a shield for surfaces that will repel, rather than kill, bacteria." The idea of repelling and not killing is important, because bacteria are harmless as long as they are prevented from landing on surfaces and forming biofilm. Trying to kill them with anti-biotics and disinfectants, such as chlorine, only builds up their resistance and leads to much-feared superbug infections that can't be treated. But Biosignal, despite having tapped into Mother Nature's genius, has a problem common to many disruptive innovations. Chlorine and other industrial biocides are plentiful and cheap commodities, so while Biosignal has a better and safer product, it doesn't yet have the volume of sales needed to bring down costs. It simply can't compete against the toxic and unsustainable chemical soups we currently cook up. New energy technologies, whether focused on efficiency or production, face the same barrier. We have a legacy infrastructure of paid-for power plants and industrial facilities that are addicted to inexpensive commodities, mainly dirty fossil fuels, that are wasteful in their consumption of energy. The big boys of industry are willing to recognize the need for change, as long as embracing it requires as little change as possible.

WANTED: SHOEHORN SOLUTIONS

Jay Harman knows this problem all too well. Take your typical household fan. The reason it often has housing around it is to compensate for inefficiency. Without housings, the blades of the fans push air both forward and outward along the tips of the blades. The air flow isn't focused; it's dispersed over a wider area and weakened as a result. Early on PAX had developed a range of fans that, because of their unique spiral design, are better at drawing in and organizing the air, a claim that has been backed up by computer simulation research conducted at Stanford University. The company estimated that the fan and motor combination it had developed was 40 percent more efficient than conventional designs, and yet it wouldn't cost more to manufacture. Also, because inefficient fans have higher air friction, they tend to be noisy. The PAX fans, by comparison, were at least 50 percent quieter. "There are about 1.5 billion of these fans sold in the United States every year," said Harman. Indeed, there are more uses for fans than just blowing air in your face on a hot day. In households alone, there are fans for air conditioners and furnaces, ceiling fans used for air circulation, fans used in humidifiers and dehumidifiers, fans in our bathroom ceilings, above the kitchen range, and inside our home computers, hair dryers, microwave ovens, and vacuum cleaners. Harman pitched the new PAX design to a leading U.S. manufacturer of domestic fans, but the company, despite acknowledging the design's superior efficiency, wouldn't bite. "They were very interested, their engineering people were interested, but at the corporate level they just weren't prepared to take it on."

Why the cold feet? Even though the new design offered their customers a better product, Harman said the top executives didn't like the fact that using the PAX design would require a retooling of their business, representing an upfront capital cost they just weren't prepared to absorb. "So in the end, we couldn't sell it even though we had a vastly superior product," said Harman. The reality is that companies invest many millions of dollars designing, testing, and certifying products and then spend millions of dollars more putting the manufacturing equipment and related systems in place to make those products. Once those systems are established, the plan is to extract as much value from them for as long as possible. This, of course, is understandable. Would you be willing to tear down the walls of a new home to put in better insulation? You're more likely to replace all your incandescent light bulbs with compact fluorescent bulbs. Incremental changes are often welcome, but in an industrial or commercial context, retooling to accommodate a new, untested technology is a much harder sell, particularly when your customers aren't complaining about your existing products or clamoring for better efficiency. As Harman said, "You've got this intractability of an industrial world that refuses to change anything."

This situation creates demand for what I call shoehorn solutions, products or technologies that are designed to fit into existing equipment, processes, or infrastructure. Shoehorn solutions are, by necessity, more incremental in what they offer because they're constrained by the inflexible environment around them. As a compromise, they betray their potential. I would argue this pressure to compromise is particularly intense as it relates to new methods for producing clean energy, which, in North America, gain more acceptance if they can be shoehorned into existing electricity and transportation fueling systems that are biased toward the use of fossil fuels. These centralized systems are controlled by large, deep-pocketed players — power generators, transmission utilities, and pipeline operators — intent on protecting trillions of dollars worth of infrastructure investments. They are the gatekeepers of energy. If you can play by their rules, then you have a better chance of being heard, and if you can bend to their ways, you have a better chance of being adopted.

For PAX it's the same story, different context. "A lot of people say, 'Yeah, what you've got there is interesting. Why don't you make a better fan that fits into our completely inefficient cooling system?'" said Harman, pointing to computer fans as one example. "We beat every computer fan in the world, and there are a couple of billion of them made every year. Trouble is there's no point taking a really good fan and sticking it into a bad environment. Computers are simply not designed to be cooled; they're designed to compute. So from the very beginning, you have to start with a design that aims to eliminate heat."

If only North Americans knew how woefully inefficient they are, how it harms their global economic competitiveness, and how using energy more efficiently is dramatically cheaper than building new energy supplies to meet wasteful demands. Consulting giant McKinsey & Company said it was puzzled to learn in 2007 how little attention was paid to energy efficiency as part of efforts to lower greenhouse-gas emissions in the United States, despite decades of public awareness campaigns, pressure from environmental groups, and a number of federal and state programs designed to support efficiency projects. A year later, it set out to identify all energy efficiency opportunities, excluding those related to transportation. It estimated that by 2020 the U.S. could reduce energy demand by 9.1 quadrillion British thermal units (BTUS), or roughly 23 percent, and eliminate 1.1 gigatons of greenhouse gas emissions annually. That's equivalent to permanently parking 200 million cars, about three-quarters of all passenger vehicles currently driving U.S. roads. "Significant and persistent barriers will need to be addressed at multiple levels to stimulate demand for energy efficiency and manage its delivery across more than 100 million buildings and literally billions of devices," the consultancy wrote. "If executed at scale, a holistic approach would yield gross energy savings worth more than $1.2 trillion, well above the $530 billion needed through 2020 for the upfront investment in efficiency measures."[5]

To what degree can biomimicry help us achieve those energy savings? It's too difficult to say precisely, but researchers at the U.K.'s University of Bath did conclude in a 2006 study that insects, plants, and animals can generally accomplish more work with less energy, at least compared to the human-engineered machines that run our economy. Their research found that most "man-made technologies" — up to 70 percent, to be specific — had to manipulate energy and often use more of it to overcome challenges, such as lifting a heavy load or dealing with the effects of extreme heat. The vast majority of creatures in nature, or what the researchers dubbed "natural machines," coped with these problems in a completely different way. "Instead, insects, plants, birds, and mammals rely on the structure and organization of their body parts and behavior," they found. "The solutions to problems are already built in." Julian Vincent, who at the time was director of the university's Centre for Biomimetic and Natural Technologies, said nature has carefully sculpted organisms into efficient machines capable of carrying out a range of engineering feats that are worth copying. "It is likely we have similar technologies to nature; it's just that we use them in a particularly unintelligent way," he said.[6] In other words, humans tend to compensate for inefficient designs by using more energy. In nature, conservation of energy is an issue of survival, so efficiency of design comes first.

BIOMIMICRY AND ENERGY

Biomimicry is an emerging field, so when surveying the landscape of energy innovators I was surprised to find so many companies that, whether intentionally or not, fit the category. Jay Harman is a seasoned veteran by comparison, but his fellow "biomimics" in the energy field who are applying lessons from nature face a wall of skepticism by going against the grain, just like PAX Scientific.

Vortex Hydro Energy, a spinoff from the University of Michigan, has developed a new kind of power-generating device that mimics how fish harness energy from flowing water, even the slowest-moving rivers. Michael Bernitsas, a professor in the university's department of marine engineering, knew that trout and salmon had an easier time swimming upstream when they harvested energy from small vortices (water swirls) that occur naturally when water flows past rocks and other objects. The fish get an extra boost by pushing off these energy-packed vortices, allowing them to reduce muscle activity and save energy. Bernitsas also knew of a naturally occurring phenomenon called vortex-induced vibration. If you stand a cylindrically shaped object in a river, you can observe small vortices being created behind the object as water flows by it. However, the vortices don't form at the same time — one on the left might emerge before the one on the right, resulting in an alternating pattern of energy pulses.

You can see this effect if you throw a fish lure with a rounded body in the water and start reeling it in. The lure will sway from side to side, just the kind of motion you want to attract a bite. But when the cylinder is a fixed object, there's no allowance for swaying. Instead you get vibration that can gather strength and cause havoc with steel risers and mooring lines, the kinds of things you might find anchoring an offshore oil platform. Bernitsas spent a good deal of his career trying to suppress how these unforgiving vibrations affected engineered structures, but one day it occurred to him that encouraging these vibrations could prove an economical way of generating electricity from slow-moving river currents, where conventional turbines just don't work. So he built a prototype of a machine called VIVACE, which stands for vortex-induced vibration for aquatic clean energy. Think of a square box about the size of a backyard shed that has no front or back wall. Inside it is a cylinder, about a foot in diameter, suspended horizontally on a mechanism that allows it to move up and down. When placed in a river, water would flow through the box and vortex-induced vibrations would cause the cylinder to bob up and down repeatedly.[7] Bernitsas' prototype captures this mechanical energy and converts it to a steady supply of direct current. The company placed a small demonstration model of VIVACE in Michigan's St. Clair River in late 2010 and has ambitious plans to launch a commercial product by 2013.[8]

Whether we're talking about Louis Michaud's tornadoes, Jay Harman's super-efficient fans and propellers, or Bernitsas' underwater vibrating cylinders, we keep coming back to swirls, spirals, and vortices. Vortices are also behind the efficiency of a new wind-turbine blade designed by WhalePower and based on the research of Frank Fish, a biology professor at West Chester University in Pennsylvania. Fish marveled at the acrobatic abilities of humpback whales, which are efficient underwater hunters capable of making sharp, tight turns, despite being up to 16 meters long and weighing more than five full-grown African elephants. He found that the key to the whale's agility is its uniquely designed flippers, which feature a row of bumps, or tubercles, along their leading edge that give them a serrated look.

Fish and research colleagues from Stanford University and the U.S. Naval Academy discovered that a humpback whale flipper has 32 percent less drag and 8 percent more lift compared to a flat-edge flipper. The basic explanation is that the water is channeled and accelerated as it flows between the tubercles, resulting in a wake of tiny energy-packed vortices that give the flipper lift. This was a remarkable conclusion for Toronto entrepreneur

Stephen Dewar, who had been following Fish's research. Dewar approached the West Chester researcher and proposed the idea of making blades for wind turbines and ceiling fans, among other bladed products, which mimic the humpback flipper.[9] WhalePower was born soon after, and its tubercle blade design has already made it into a ceiling fan manufactured by Envira-North Systems, the largest supplier of industrial ceiling fans in Canada. Envira-North says its whale-inspired fan, as odd-looking as it may be, is at least 20 percent more efficient at moving air and also dramatically quieter. In its marketing material, the company touts the fan as having benefited from "a million years of field tests." The tubercle blade design's quietness alone could prove highly beneficial for manufacturers of wind turbines, considering noise pollution is one of the most frequent complaints targeted at industrial wind farms. But so far the Vestas and General Electrics of the world haven't shown much interest and appear in no rush to change their systems to accommodate a new design.

Nature can certainly teach us how to move through water and air more efficiently, but it can also offer tips on how to manage our consumption of energy. REGEN Energy, another Toronto company, has developed a way to control electricity use in large buildings by using the same "swarm logic" that bees use to work through problems. An individual bee has a brain the size of a sesame seed and on its own is just another dumb bug. But get a thousand bees together, and they connect as part of a larger, more intelligent collective, one where no particular bee is calling the shots. This magical approach to collective action isn't just a bee trick: ants exhibit this self-organizing behavior as do crickets, and in functioning this way, they demonstrate a remarkable ability to coordinate their activities. Humans generally operate in the opposite way. We've grown accustomed to command-and-control systems, where decision making is centralized and flows from the top down.

The top-down, central command model is inefficient with respect to how devices in a network consume electricity, thought REGEN founder Mark Kerbel, who in 2005 turned to swarm logic as a way to "take the bloat off the grid." He and his business partner, Roman Kulyk, invented a proprietary algorithm that they embedded in tiny wireless controllers attached to fans, air conditioners, pumps, refrigerators, and other building equipment designed to cycle on and off. Building owners often pay a hefty utility premium when their power demands spike, so it's in their interest to make sure their appliances don't all cycle on at the same time. Connecting all appliances to a central system would be too expensive and complex. REGEN's wireless controllers, on the other hand, act like bees. They constantly send simple messages to each other to determine who's off and who's on, and they'll make their own decisions based on the information received. There might be a few dozen controllers in a building working together as a single "hive," and through their frequent exchange of messages they manage to smooth out the electricity load in a building and eliminate costly power spikes.[10]

Down the road, Kerbel sees the potential of using his devices to assure the orderly charging of electric cars as major automakers bring more zero-emission vehicles to market. As Kerbel explained, "As soon as you come home, for example, your plug-in Prius would start talking to other plug-in cars in the neighborhood and they would all come up with a plan to charge evenly without overwhelming the grid." It may prove to be the best way for local utilities to prevent blackouts in communities where plug-in electric vehicles represent a rising share of the total automotive stock.[11] But embracing systems based on swarm logic also requires a leap of faith. It's not easy giving up control and putting trust in a collective of wireless devices. It may work in nature, but no bee ever got fired from its job by putting its faith in the hive.

Janine Benyus is directly involved in a bio-inspired energy venture, OneSun Solar. The stealthy California startup, headed by prominent environmentalist and entrepreneur Paul Hawken, is developing "extremely inexpensive" solar cells that function like the leaf of a plant.[12] They don't use toxic metals like cadmium or rely on harmful gases, such as silane, during manufacturing. The technology mimics photosynthesis and, according to

Benyus, will offer an energy return of 200 to 1 and make solar energy more affordable than coal or nuclear power. "We decided in the scoping phase of our design that we should manufacture without heat, that the product would self-assemble, that it would rely on water-based processes, and it wouldn't rely on rare Earth metals," said Benyus, who is a fan of solar power but is worried about the landfill legacy current solar PV technologies based on toxic metals and processes will leave behind. "We don't want to have to build a Yucca Mountain for all our solar waste."[13]

FAITH IN NATURE

The above ideas and technologies are truly inspiring, but try walking into a board meeting or a room full of old-school engineers and mentioning words like biomimicry, Mother Nature, and even *pax*, which is the Latin word for peace. If heads don't roll, many eyes certainly will. Disdain for his ideas is exactly what Harman often experienced during the early days of PAX Scientific. He recalled presentations he did in 2002 to a group of about 10 engineers at United Technologies, a multibillion-dollar industrial giant that makes everything from air conditioners to helicopters. "At the end of the presentation, the head of engineering stood up and said, 'This is bullshit! You've been wasting our time,' and he walked out." Another pitch a couple of years later, this time for the U.S. Office of Naval Research, invited a similar response. "This is nonsense," said one naval engineer. "Technology shits on nature!" Harman laughed when recalling that story, only because he knows that no engineer, no scientist in the world has yet demonstrated a single situation where human design is more efficient than the natural designs around them. "Humans don't even come close, in every single case," he said. "There's no scientist on Earth that can explain the flight efficiency of a single insect."

Attitudes are changing, at least from Harman's perspective. Around 2006, when Al Gore's award-winning movie *An Inconvenient Truth* was released, the public and businesses began waking up to the threat of climate change and the role being played by fossil-fuel use and wasteful energy practices. Then on July 11, 2008, the price for a barrel of oil hit a record $147.27, a dramatic milestone when one considers that nine years earlier, oil could be purchased for less than $20 a barrel. The cost of doing business was suddenly much more expensive, and what became an energy crisis shined a light on the value of energy efficiency. Consumers, meanwhile, began dumping their gas-guzzling suvs and opting instead for more fuel-efficient cars, or they simply drove less.[14] Oil prices did plunge back down to $30 a barrel as the global economic recession took hold and demand for oil began to plummet, but this was a temporary retreat. Through 2010, for example, oil hovered consistently between $70 and $80 a barrel, two or three times more expensive than oil prices during the 1990s. For much of 2011, oil has been above $100 a barrel. Surprisingly, these high prices were sustained during a period of relatively weak demand, indicating that oil prices, while they may bounce up and down over the short term, are on an upward trajectory over the long run. The cost of electricity, as pointed out in chapter 2, has also been rising steeply over the past decade, with average retail prices in the United States up nearly 50 percent. The best way to cope with this reality is to embrace efficiency. "We must re-engineer our lives to adapt to a world of growing energy scarcity," according to economist Jeff Rubin. "And that means learning to live using less energy."[15]

"Until that energy crisis, you couldn't give energy efficiency away in the United States," said Harman, adding that compa-nies are now beginning to approach PAX because they know that energy efficiency is strategic to their survival. "Now we're in a very fortunate position. We're in the right place at the right time." When the recession hit in late 2007, the U.S. government reacted by passing the American Recovery and Reinvestment Act of 2009, a huge economic stimulus package aimed at re-firing the economy. It recognized that investments in energy efficiency weren't just a path to job creation; they would also prepare consumers and businesses for higher energy prices and help to boost U.S. industrial productivity. Many billions of dollars were set aside in the stimulus for energy efficiency initiatives, including $4.5 billion targeting federal government buildings, $600 million aimed at industrial facilities and commercial buildings, and $300 million in rebates toward the purchase of energy-efficient appliances.

Ask U.S. energy secretary Steven Chu why energy efficiency has been underappreciated in the past, and he'll tell you that the market simply failed to value it. Inertia, ignorance, and lack of financing for projects have been a big part of the problem, as has the perception that retrofitting systems and switching out equipment is too much of a hassle and inconvenience. But in recent years, the U.S. Department of Energy has made a genuine effort to overcome these barriers. Commenting on the importance of energy efficiency, Chu wrote in 2010, "Regardless of what the skeptics may think, there are indeed $20 bills lying on the ground all around us. We only need the will — and the ways — to pick them up."[16] The commitment from the U.S. government has started to benefit PAX, which through its various spinoff companies has since received millions of dollars in development grants. It may still be an underdog compared to more established and better understood energy-efficiency technologies on the market, but the initial pushback the company experienced during its early years has been replaced with some welcome pull. "The government is starting to support biomimicry, I'm happy to report," said Harman. "We hung on by our toenails and fingernails for years. It was a serious struggle if you ask any of the folks in our companies, but now we're actually breathing; the businesses are breathing on their own. We're out of intensive care."

PAX is a family of companies now. Engineering and research activities remain within PAX Scientific, but subsidiaries include PAX Streamline (power generation, propulsion, heat/air/ventilation systems), PAX Mixer (mixing technologies), and PAX Water (systems for handling water and wastewater). As well, Paul Hawken runs a completely separate company called PaxFan LLC, which, as part of a licensing arrangement, markets PAX

Scientific's fan technology for use in computer equipment, automobiles, industrial equipment, and domestic appliances. Though business is better, Harman doesn't fool himself into thinking that his work will get any easier. He knows that for PAX to have the impact he envisions, and for biomimicry in general to gain mainstream acceptance, it has to show that changing the way we consume energy by learning from natural forms and processes is, in the end, smart business. "I didn't leave the fisheries and wildlife department because I wanted to become a corporate person, because I absolutely don't — I'm a bushman," he told me. "But it became pretty apparent to me that things in this world aren't going to change unless you can show it's profitable to change them. Let's face it, the world is run by corporations, and corporations are run by bean counters. So if you're going to do biomimicry, that's great, you might have a better solution. But it's only going to mean anything if it's more profitable."

The road from conception to proof-of-profitability is a long one that, far too often, ends prematurely and unexpectedly. Breakthrough technologies don't always break through. Money runs out. The will to go on evaporates. Passion, hope, and good intentions are replaced with cynicism. Having a superior technology alone, as we've seen so far, isn't enough to complete the journey from discovery and development to demonstration and commercialization. Yet Harman is an example of how — despite the pushback, the skepticism, and the hard-held belief by some that human-conceived technologies "shit on nature" — persistence and patience *can* pay off. His Lily Impeller and other spiral-shaped fans and turbines are beginning to move more of the planet's fluids and gases. There may even come a tipping point, when the spiral products based on PAX-developed algorithms become a global standard. But there will be no overnight disruption here, no butterfly suddenly emerging from its chrysalis. Infrastructure is slow to change. So, too, are people and policies.

Whatever humanity does, of course, Mother Nature always wins in the end. We are but a brief entry in her datebook. So if you can't beat her, join her, or at least copy her — so say the biomimics who are leading the way. Some innovators have a different approach. Why rip off Mother Nature when you can get her to do most of the work for you? That's what Paul Woods is doing with his Florida-based startup Algenol, the subject of the next chapter. Woods is an energy farmer, growing algae that can supply a seemingly endless flow of clean-burning ethanol fuel. Twenty-five years ago, the big oil companies laughed at the idea. Not anymore.

Notes:

1 "Challenges facing green business start-ups," Mongabay.com. October 20, 2008.

2 Philip S. Callahan, *Tuning In To Nature: Solar Energy, Infrared Radiation and the Insect Communication System* (Old Greenwich, CT: Camelot, 1975), 94.

3 Janine Benyus, *Biomimicry: Innovation Inspired by Nature* (New York: HarperCollins, 1997), 2.

4 The Biomimicry Institute has set up an online database called AskNature.org, which allows inventors (designers, engineers, architects, etc.) to better connect what they're trying to do with what nature has to offer. For example, if you want to filter salt from water you type that in and the database will come up with examples showing how nature performs this trick. The goal is to encourage inventors to explore biomimicry and adapt natural mechanisms into their designs.

5 McKinsey & Company, "Unlocking Energy Efficiency in the U.S. Economy." July 2009.

6 Julian F.V. Vincent, et al., "Biomimetics: It's Practice and Theory," *Journal of the Royal Society.* Paper accepted March 27, 2006. See also: "Copying Nature Could Save Us Energy, Study Shows," press release, University of Bath. May 9, 2006.

7 A video showing this up-and-down motion can be seen at http://www.vortexhydro energy.com.

8 Tyler Hamilton, "A New Twist on Hydropower," *Technology Review.* December 3, 2008. http://www.technologyreview.com/energy/21749/.

9 Tyler Hamilton, "A Whale of a Tale," *Toronto Star.* May 14, 2007. See also Tyler Hamilton, "Whale-inspired Wind Turbines," *Technology Review.* March 6, 2008. http://www.technologyreview.com/energy/20379/.

10 Tyler Hamilton, "Lots of Buzz Surrounding This Company," *Toronto Star.* September 3, 2007.

11 Tyler Hamilton, "Electric Cars Are Coming, But Can California Take Charge?" *Toronto Star.* July 12, 2010.

12 Benyus and Hawken are both on PAX Scientific's advisory board along with Amory Lovins.

13 Yucca Mountain is a mountain in Nevada that has been under study for 32 years as the long-term storage site for radioactive waste from U.S. nuclear power plants and other sources. In 2009, the administration of U.S. President Barack Obama announced his cancellation of the Yucca project, though discussion of the project remains very much alive.

14 "Peak Demand — U.S. Gasoline Demand Likely Peaked in 2007," press release, IHS Cambridge Energy Research Associates. June 19, 2008.

15 Jeff Rubin, *Why Your World Is About to Get a Whole Lot Smaller* (Toronto: Random House Canada, 2009), 23.

16 Steven Chu, "Energy Efficiency: Achieving the Potential, Realizing the Savings," *Huffington Post*. March 16, 2010. (His commentary originally appeared in "Energy Vision 2010: Towards a More Energy Efficiency World," a report by the World Economic Forum and IHS Cambridge Energy Research Associates.)

CHAPTER 5

Not Your Average Pond Scum

Making Fuel Refineries Out of Algae

"As a group, algae may be the only photo-synthetic organism capable of producing enough biofuel to meet transportation fuel demands."

— Dr. Timothy Devarenne, Texas A&M University

Tar balls and crude patties. That's what came to mind while driving down a sunny stretch of road in West Palm Beach, Florida's self-proclaimed "city of unsurpassed beauty." It was June 22, 2010, and while my trip dropped me on the side of Florida left untouched by BP's rapidly expanding oil slick in the Gulf of Mexico, I knew that other parts of the Sunshine State weren't so lucky. Still gushing after two devastating months, the spill that had left black gooey streaks and oil puddles along the coasts of Louisiana, Mississippi, and Alabama had begun to tarnish Florida's northwest Panhandle. Pensacola Beach, usually bustling with tourists at this time of the year, was uncharacteristically sedate, a heartbreaking situation in a state already in the grips of a housing crisis.

Minutes from my destination, I turned onto a country road and drove for two or three kilometers before realizing I had gone too far. Throwing the rental car in reverse, I backed up 50 meters and finally spotted the driveway that took me to an inconspicuously located outdoor test facility operated by Algenol Biofuels. Long, narrow rows of knee-high greenhouse structures stood next to a cluster of office trailers, all of it hidden to the outside world by walls of wind-blown palm trees. Marine biologist Frank Jochem, director of the five-acre facility, met me at my car and we walked to a trailer where I was introduced to Harlan Miller, a photosynthesis expert who joined the company in 2008. As the three of us waited for Paul Woods, the Canadian-born founder and chief executive of Algenol, the BP oil spill that had occupied my mind during the drive became the topic of conversation. Jochem, speaking with a German accent, shook his head in obvious disgust. "I don't know what's worse, all of the leaked oil or the chemicals they're using to fight the slick."

Disturbing times, yes, but it makes their work at Algenol that much more meaningful. Jochem and Miller are part of a growing team that is pushing toward a future where feeding our fuel and chemical addictions doesn't mean drilling deeper offshore or relying on sticky bitumen hastily extracted from the tarry sands of northern Alberta. The company has developed and patented a low-cost method for producing clean-burning ethanol fuel from genetically enhanced blue-green algae, otherwise known as cyanobacteria, a photosynthetic cousin of algae that most people would lump into the category of pond scum.[1] That's right, as strange as it sounds, within the next decade the car you drive could be powered by fuel made from the guts, sweat, or excrement of microscopic organisms that would rather be hanging out in your fish tank. The seed of Algenol's creation was planted in the early 1980s when Woods, then a 22-year-old undergraduate studying genetics at the University of Western Ontario, learned that some species loosely labelled "microalgae" can naturally produce ethanol when subject to certain environmental conditions.[2] Woods wondered if these organisms could be genetically engineered to maximize ethanol production and, after convincing himself they could, he pursued the idea with vigor.

Nearly three decades later, what began as a student's curiosity — a "glorified hobby," as Woods described it — is now a thriving startup with more than 100 employees, $100 million plus in financial backing and Fortune 500 partnerships that include chemical giant Dow Chemical, industrial gas supplier Linde Group, and oil-refiner Valero Energy. Getting from there to here wasn't a straight line. There were many bends, bumps, and detours along Wood's journey, which is still far from over and not likely to get any smoother. But if Algenol plays its cards right — and like all of the stories in this book that's a big *if* — the company's algae-fuel process could help make deep-sea drilling and other high-risk oil projects appear like foolish, uneconomic pursuits.

RECYCLING CO₂

Let's begin with a little primer on ethanol. Also known as grain alcohol, ethanol is a renewable fuel that has the potential to displace up to 85 percent of the gasoline used in North American passenger vehicles.[3] It can also replace fossil fuels in the production of ethylene, the basic chemical feedstock for making many types of plastics. The reason ethanol is "renewable" is because we don't make it from the skeletons locked away in Earth's closet, like fossil fuels made up mostly of dead microorganisms that have accumulated over tens of millions of years. Instead, ethanol is made from plants that are part of a natural closed-loop cycle of growth, death, and decomposition on the Earth's surface. Burning ethanol releases carbon dioxide but in a carbon-neutral way, meaning the carbon in the CO_2 emitted during combustion is re-consumed by the growth of new plant life. In theory, at least, no new sources of carbon are introduced to the cycle; the carbon is essentially recycled. The release of CO_2 from burning fossil fuels such as oil and coal, on the other hand, is carbon-positive because it adds carbon to the atmosphere that was previously held in deep storage within the Earth's crust. The more we take from underground and burn at the surface, the higher the concentration of sun-trapping CO_2 in the atmosphere and the greater the greenhouse-gas effect playing havoc with climate systems. "The bottom line is that atmospheric carbon dioxide acts as a thermostat in regulating the temperature of the Earth," according to NASA climate scientist Andrew Lacis. "The rapid increase in atmospheric carbon dioxide due to human industrial activity is therefore setting the course for continued global warming."[4] If we are to keep climate change under control, we must leave as much of that "ancient" carbon permanently underground. We need to keep the microscopic skeletons in Earth's Paleozoic closet locked up, and we need to throw away the key.

But ethanol isn't without controversy, which mostly concerns how it's made and what's used to make it. In North America, virtually all of the ethanol that's produced is made by fermenting the starches in corn, a crop that is heavily subsidized and energy-intensive to grow. Most studies suggest that, based on the entire production lifecycle of corn-derived ethanol, the energy required to plant, grow, harvest, transport, and ultimately process the corn into ethanol is almost as much as the energy you get from the ethanol itself. With such a poor energy balance, and because the energy inputs are likely to be fossil fuel in origin, there are serious doubts in the scientific community about the emission-reduction benefits and sustainability of making ethanol from corn. Complicating matters is that corn is food for both humans and animals. As more of the world's corn crops are used to make fuel, there is widespread concern, and already plenty of evidence, that food and animal feed prices will rise.[5]

Brazilians have been successful at making ethanol from sugarcane, using a process that returns several units of energy for every unit of energy put in. Sugarcane, however, is a tropical crop that for obvious reasons isn't big business in North America. Besides, dense carbon-rich forests in Brazil are being clear-cut to make more room for sugarcane fields. This practice results in more greenhouse-gas emissions, not less, and defeats the purpose of transitioning to ethanol. To move beyond these controversies, U.S. and Canadian efforts at improving ethanol's energy balance have focused largely on technologies that can convert a wider range of plant materials — industrial wood and paper waste, forest slash, agricultural residues such as corn stover and wheat straw, and dedicated crops such as switchgrass — into cellulosic ethanol, meaning no more dependence on food crops such as corn and sugarcane. Cellulose gives structure to the cell walls of plants and makes up the majority of all plant matter. Sometimes called roughage, it's the fiber in plants that we find difficult to digest. But here's the thing: cellulose has a lot of sugar locked up inside it; the challenge is getting it out. So those scientists making cellulosic ethanol must rely on the use of special enzymes, such as those found in the guts of termites or used by certain types of fungi, which have an easier time than humans in liberating the sugars in cellulose. Another approach is to use the right balance of heat, pressure, and oxygen to directly convert or "gasify" the biomass into synthetic gas, which is then fed to special microbes or some other catalyst that turns the gas into liquid ethanol.

Methods vary, as do results, but the return on energy is generally better. Making cellulosic ethanol from fast-growing switchgrass, for example, is believed to return 3.5 to 7 units of energy for every unit consumed. It can also lower greenhouse-gas emissions by 94 percent compared to gasoline.[6] The overall goal with cellulosic ethanol is to not compete with food crops or for land that is ideal for growing food (or which is already heavily forested). The aim is to produce more fuel using less space, reduce reliance on fresh water, and constantly improve the energy output of the fuel relative to the energy inputs required to make it. The underlying problem is that cellulosic ethanol, compared to corn ethanol, remains expensive to make even after many years of research and demonstration. "Today's cellulosic ethanol is competitive with the petrol it is supposed to displace, but only when the price of crude oil reaches $120 a barrel," according to an October 2010 report in *The Economist*.

That's where Algenol hopes to demonstrate a significant edge. It claims it can produce ethanol from blue-green algae that is competitive with oil priced as low as $30 a barrel and it has an energy return of 5 to 1. And Algenol can do it using marginal land, dramatically less space, and zero fresh water. All it needs is a steady supply of sunshine, carbon dioxide, and seawater — all of which are plentiful — and some nutrients (phosphorous and nitrogen) for the algae to munch on. Animal manure, for example, is nice snack food for most algae. In return, Algenol says it will produce a steady flow of renewable ethanol fuel for less than $1 per gallon that displaces the use of gasoline and petroleum-based chemicals. As an added bonus, the company says that for every gallon of ethanol it produces, it also produces a gallon of fresh water. That feature alone could prove compelling for regions, such as Australia or the U.S. Southwest, plagued by chronic drought, and in many African countries where freshwater supplies are limited.

REFINERY IN A CELL

In his 20s, Paul Woods was so passionate about the algae-to-ethanol concept he had thought up in university that he spent five years after graduation trying to build support for it. He was determined to be an entrepreneur, to not get trapped in a safe, nine-to-five job like his parents had during his middle-class upbringing. His dad was a computer engineer and his mom an insurance agent, and their advice to him emphasized carefulness and caution.[7] His instincts, on the other hand, told him to take risks and trust his gut. Blinded by inspiration, he traveled around Canada during the mid-1980s meeting with officials from oil and gas companies in a gutsy yet naive effort to win them over. "Honest to god, I thought the merit of the idea was good enough, was so fantastic, so simple and elegant, that somebody out there would get it," Woods, now in his late 40s, recalled while we sat huddled, along with Jochem and Miller, in an office trailer at the West Palm facility. "I went to companies such as Suncor, or Sunoco at the time, and everybody hated it. They just hated it. They thought I was out of my goddamn mind." At the time, who wouldn't?

Like PAX Scientific's Jay Harman, Woods isn't a suit. He showed up for our meeting in shorts, sandals, and a collared short-sleeve shirt with a funky paisley design. With red, nearly shoulder-length hair, he has a laid-back demeanor that suggests he could have been a party animal in university and may still be today. He's professional but in a fun way, stern when he needs to be but willing (at least in my presence) to chat and joke as if sharing a drink with friends. At one point in the conversation, the chatter turns to hockey, a sure sign that two Canadians are in the room.

Looking back at his early attempts to woo the oil companies, Woods realized his timing couldn't have been worse. Climate change was non-existent as a political issue, peak oil concerns remained at the fringe, and an oil glut following the 1970s energy crisis meant the price of crude in the mid-1980s was less than half of what it cost at the beginning of that decade. He might as well have been selling tofu to cattle ranchers. Woods turned to fine art through the latter half of the '80s; he owned and operated a small art gallery in his hometown of Markham, Ontario, and it was there that he got some sound advice from a supplier and friend. "He said, 'Look, Paul, if you're ever going to do this, you can't rely on anyone else. You're going to have to do it yourself. You're going to have to raise your own money.'" Re-energized, Woods discussed his algae fuel idea with family and friends, even customers of his art gallery, and by the late 1980s had managed to raise about $200,000 by selling $1,000 shares to 70 individual investors.

It was shortly afterward that Woods met John Coleman, a plant biologist working as a professor in the cell and systems biology department at the University of Toronto. "He came to the university and picked me out from a local expert list we have here," Coleman remembered. "Paul basically walked into my lab and said, 'Do you think this is a possibility?' I sat down and thought about it for a while, and we started coming up with some more ideas." They knew many types of blue-green algae (cyano-bacteria) made ethanol naturally but not in a way that is linked to photosynthesis. Instead, ethanol is produced in small quantities when the cyanobacteria are under stress — starved, without oxygen, and in the dark. The organisms thrive when exposed to sunlight, carbon dioxide, and nutrients, which through photosynthesis produce sugars that are directly metabolized into carbohydrates, such as glycogen. The trick for Coleman was to get the cyanobacteria to convert sugar into ethanol when they weren't under stress, to have them produce ethanol under ideal conditions before the sugars are turned into carbohydrates. That meant identifying the enzymes responsible for ethanol production and finding a way to switch them on when they would normally be off. The second trick was to significantly boost the quantity of ethanol produced by the organism. The answer here was to add in some genetic material, an enzyme, from another type of bacteria, such as *Zymomonas mobilis*, a hardy organism that's used instead of yeast to make alcoholic beverages in many countries. By genetically modifying the metabolic priorities of cyanobacteria and enhancing them with the help of a foreign enzyme, Woods and Coleman knew they could turn blue-green algae into highly efficient microscopic refiners of ethanol.

What followed through the 1990s was an on-and-off research effort led by Coleman and financed by Woods. Funding for the first two years came from the $200,000 that Woods had raised from friends and family, as well as an additional $60,000 grant from the Ontario government. After that money ran out, Woods self-financed the research when he could. And fortunately for him, he could; Woods is the type of guy who keeps many pokers in the fire. The same year he hooked up with Coleman, he also started an energy retail business that sought to take advantage of deregulation in Ontario's natural gas market. In 1989, he formed Alliance Gas Management, which purchased large quantities of natural gas at a volume discount and resold it to customers through fixed-rate contracts. Reselling energy services such as natural gas and electricity is common today, but back then Woods was a pioneer. He was told by energy industry experts at the time that the idea would never work. It didn't sway him. "I made up brochures on my home computer, which was a little Macintosh SE, and a friend of mine printed them out," he told me. He mass-mailed the brochures and, to his delight, customers began signing up. "It worked like a charm."

Eight years later, Alliance Gas had an initial public offering that raised $80 million and in 1999 the company, boasting nearly half a million customers and more than $100 million in revenues, was acquired by Centrica subsidiary Direct Energy, the largest energy retailer in North America. The same year as the public share offering, Woods founded United Gas Management as part of a plan to enter the U.S. market by replicating the success he had in Ontario. "We ended up in nine U.S. jurisdictions," said Woods, who in just three years grew United Gas to 240,000 customers. He sold his shares in 2000, however, after the credit-squeezed company ran into financial troubles. A seasoned executive who had proven his ability to manage and grow companies, Woods was glad to be out. "It was a nasty business," he recalled. "The utilities just viewed it as us stealing their customers, and they did whatever it took — lied, cheated — to get us out of their territories." Rather than throw himself back into that ring, Woods, who at 38 years old had accumulated significant wealth, decided to throw himself into early retirement.

Throughout all of this, Coleman kept at his algae-to-ethanol research and Woods continued funding it out of pocket, to the tune of about $700,000. The retired life was good for Woods, now a resident of West Palm Beach. So good, in fact, that he was in no rush to do anything but enjoy it. "I didn't work for six years," he told me as we drove for lunch in his white Rolls-Royce. "I just ate and drank and traveled." As Woods' biologist-for-hire, Coleman had made significant progress with his research, but Woods knew the market wasn't yet ready for an algae-to-ethanol venture. Oil could be purchased for $20 a barrel in 2000, the year of Woods' retirement, so the economics of making ethanol from algae didn't make sense; pursuing the market at the time would have been suicidal. The turning point didn't come until 2005, when oil prices burst well past $50 a barrel in the face of rising demand from China and disruptions to U.S. production and refinery capacity in the aftermath of Hurricane Katrina. Woods' mother called him one day and, during the conversation, began complaining about the impact all that expensive oil was having on gasoline prices.[8] This was the tug Woods needed. He was back to business and better equipped to carry his algae-to-ethanol concept forward.

ALGAE ORGY

It's worth mentioning that using algae to make biofuels isn't biomimicry. Scientists aren't trying to learn or copy from algae. Rather, they're determined to get algae to do the work for us by isolating, tuning, and tampering with their genetic makeup. Janine Benyus would call it bio-assistance — using, training, or modifying an organism to do what we need it to do. Using eggs to incubate flu vaccines is an example, as is the use of bacteria for the production of insulin. You could also call it bio-enslavement. It was in the 1930s that the idea of turning algae into microscopic power plants first captured the imagination of scientists. It began with a German researcher named Hans Gaffron, who after fleeing the Nazi party settled into a career at the University of Chicago. There, in 1939, he discovered that algae often produce hydrogen, under then-unknown conditions.[9] Back in Germany, scientists were also experimenting during and after World War II on the ability of algae to produce oil, which they observed could represent up to 70 percent of an organism's dried weight.[10]

Decades passed with only sporadic research until, in 1978, the administration of U.S. President Jimmy Carter established the Solar Energy Research Institute, the predecessor to today's National Renewable Energy Laboratory (NREL). The institute was a response to the 1970s energy crisis and created within it was the Aquatic Species Program, which studied the use of algae and other aquatic plants as a resource for energy production. The program's first four years focused, like Gaffron did 40 years earlier, on hydrogen production, but in 1982 research began shifting to the extraction of oils for the purpose of making biodiesel. The program was small, had limited funding, and, though its scientific contributions were substantial, fell victim to the Clinton administration's deficit-fighting of the mid-1990s. In 1995, the U.S. Department of Energy pulled the plug on algae research and redirected most of what resources remained toward the study of ethanol, though the focus was on cellulosic ethanol and not the production of ethanol from algae.

Researchers involved with the program seemed to lament the decision. They knew that algae had several advantages over other approaches to biofuel production. Unlike corn, these simple-cell organisms don't rely on the use of arable land and don't necessarily use up freshwater resources. Certain strains can thrive in seawater and wastewater. They also reproduce at an astounding rate and are capable of doubling their mass in just a few hours, blowing away the productivity of dedicated non-food crops such as switchgrass. Algae are productive little creatures that, when exposed to sunlight, are unrivaled in their ability to gobble up carbon dioxide, even relatively dilute sources from a fossil-fueled power plant. Theoretically, these high achievers of photosynthesis can produce 10 to 100 times more oil per hectare than traditional crops such as corn.[11] Outside the ideal conditions of a government laboratory, however, the potential of economically produced algae oil remained a best guess.

"Algae are not nearly as well understood as other organisms that have found a role in today's biotechnology industry. This is part of what makes our program so valuable," the NREL scientists wrote in a 1998 review of their work. "Much of the work done over the past two decades represents genuine additions to the scientific literature. The limited size of the scientific community involved in this work also makes it more difficult, and sometimes slower, compared to the progress seen with more conventional organisms." They concluded then that the cost of producing biodiesel from algae was still too rich to compete against oil, and that the ability to make enough of it to have an impact was still in question. Still, encouraged by advancements in genetic engineering, the scientists held out hope: "This report should be seen not as an ending, but as a beginning. When the time is right, we fully expected to see renewed interest in algae as a source of fuels and other chemicals."[12]

They got that last part right. It wasn't just Paul Woods who anticipated the hat-trick of rising oil prices, rising CO_2 emissions, and, thanks to terrorism and political instability, rising concern over energy security. One of the first companies, if not *the* first, to emerge on the algae-to-biofuel scene was GreenFuel Technologies, founded in 2001 by a 34-year-old chemical engineer named Isaac Berzin. GreenFuel was unique, ambitious, and willing to make the kind of bold claims that create media darlings. Becoming one of the best-funded companies in the game, GreenFuel raised awareness of the potential of algae fuels and fired up the imaginations of scientists, entrepreneurs, and investors who were watching its story unfold. New algae-fuel ventures with sci-fi sounding names soon began to sprout. We saw the emergence of San Francisco–based Solazyme and Seambiotic of Israel in 2003, while LiveFuels of San Carlos, California, and Solix Biofuels of Fort Collins, Colorado, made their debuts in 2006, along with Algenol. San Diego's Sapphire Energy, which counts billionaire Bill Gates as an investor, arrived on the scene in 2007. By the end of 2010, the market was flooded with startups claiming the ability to solve the world's energy woes by turning algae, that slimy green stuff that builds up in our fish tanks and dog bowls, into renewable fuel.

There was plenty of hype, sure, but collectively there was more talk than results. The Energy Biosciences Institute at the University of California–Berkeley remarked in an October 2010 report that algae biofuel production had failed to move beyond small-scale pilot plants, despite more than 100 companies in the United States alone pursuing the market. "The total output from all experimental facilities over the past year was only a few tons of biomass and less than 100 gallons of actual algae oil, if that much," it concluded.[13] That works out to less than 379 liters, versus the 758 *billion* liters of transportation fuels produced in the United States each year. Understandably, there has been some backlash to the hype. Estimating the existence of about 200 algae-energy companies worldwide by the end of 2009, *Mother Jones* writer Emily Waltz was quick to pounce in a story titled "Algae Energy Orgy" that labeled the majority of ventures as "scum artists" with empty promises. "Some companies have promised impossible amounts of oil based on speculation, raising millions from unwitting investors," wrote Waltz, who seemed gleeful in pointing out the demise of GreenFuel in early 2009 as the first of many failures to likely come.[14] Indeed, there's no saying Algenol won't be one of them.

ETHANOL ENDGAME

If out-of-lab production of algae biofuels has been less than stellar, at least there have been major advancements around the growing, harvesting, and processing of algae for fuel. John Sheehan, a researcher at the University of Minnesota who was involved with early algae research at NREL, figures that nearly 12,000 liters of algae oil can be produced for every hectare of land each year, compared to just under 4,000 liters of ethanol made from corn and only 3,300 liters of cellulosic ethanol from biomass. He estimates that before 2020, the per-hectare output of algae oil could shoot up to 45,000 liters a year, compared to 14,000 for cellulosic ethanol and less than 6,000 for corn ethanol.[15] Clearly, the upside sits with the green slime team. Indeed, a 2011 study by Pacific Northwest National Laboratory, a U.S. Department of Energy lab, argued that biofuels made from U.S. grown algae could displace 17 percent of the country's oil imports — roughly equating to 21 billion gallons of algal oil.[16] Personally, I think that's low-balling its potential.

But results depend on methods, and not all algae are created equally. Some groups (like LiveFuels, Seambiotics, and Sapphire Energy) grow algae in shallow open ponds that are exposed to the elements; fewer (Innoventures Canada) use covered ponds where temperature and growing conditions are easier to control, while others (Solix) prefer smaller enclosed spaces called photobioreactors, or PBRs, that are designed to soak up the sun and receive controlled batches of CO_2. Most companies harvest the critters when they're full grown, crush or grind them up, and then squeeze out their internal oils; some (Pond Biofuels) are happy to just dry the algae and burn them directly for their energy, whether to produce heat or generate electricity. Some companies (LiveFuels) prefer autotrophic algae strains that rely on sunlight and carbon dioxide to grow, while others (Solazyme) go with heterotrophic algae that don't rely on photosynthesis but instead thrive in the dark when fed a consistent meal of sugar, wastewater, or some other carbon-rich nutrients.

There are trade-offs with each approach. Going with an open pond is far cheaper but sacrifices the control and high algae productivity one gets with more expensive and complicated PBRs. Getting at and extracting algae oil requires multiple costly steps but can lead to higher-value fuels and chemicals (and leftover proteins that can be sold as animal feed), while the more direct option of burning the algae results in lower-value heat energy. Autotrophic algae grow by consuming CO_2, but their need for sunlight creates geographic limitations and engineering challenges to make sure sun exposure is evenly distributed and maximized, while heterotrophic algae can be cultivated practically anywhere, in dark enclosed tanks 24 hours a day. But absent their ability to consume CO_2 through photosynthesis, these heterotrophs must be directly fed carbon-based nutrients, such as sugar. This adds cost and eliminates one of the major attractions of algae fuels, which is the opportunity to recycle carbon from fossil-fueled power plants. Under all of these scenarios, algae must be bred, genetically manipulated, and ultimately customized to the environment they're placed in. So where does Algenol fit in?

Algae can be broken down into three main components: oil, carbohydrates, and protein. Algenol is exceptional in the market for its decision to breed algae that are robust refiners of ethanol,

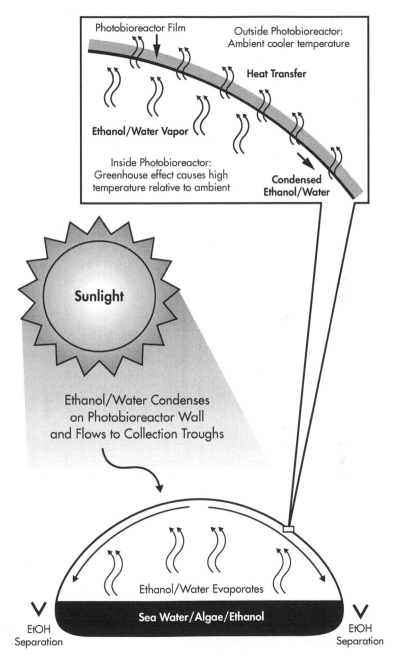

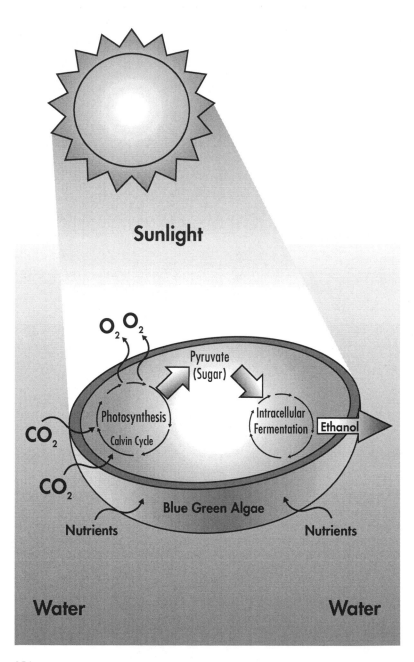

not one time sources of unrefined oil. I say "exceptional" because there appears to be no other algae companies at the time of writing that are pursuing the same ethanol endgame, and most major studies of algae biofuel — from the early studies at NREL to recent reports from the Energy Biosciences Institute — specifically exclude commentary and analysis related to ethanol production, which has been largely ignored in favor of algae oil research. The result is that Algenol has had to do some pioneering research to get it where it is today, leaving it somewhat alone in the race to bring algae biofuels to market.

Hearing John Coleman talk about it, the focus on ethanol makes a lot of sense. "Ethanol is almost infinitely mobile in an algae cell," he told me. "It essentially leaks out of the microorganism after it has been produced in the cell. It evaporates into a gaseous phase." This is immensely important. Getting oil from algae usually means constantly growing, fattening up, and harvesting the organisms, then going through a number of costly steps to extract what is basically vegetable oil that must be further refined into end products, such as diesel or jet fuel. As one scientist told me, "We don't grind up cows to get their milk, so why should we do that with algae?" Coleman explained that the secretion of ethanol out of cyanobacteria cells is comparable to how plants sweat water, a process referred to as transpiration. In both cases, it just happens — no external energy is required to stimulate or sustain the process. This means the algae are left alone to soak up the sun and CO_2 as they produce ethanol; they don't have to be harvested and killed to extract the ethanol as is the case with most algae-to-oil processes. So while the vast majority of algae biofuel ventures must manage the constant replenishing and harvesting of their algae stock, Algenol keeps its algae alive and productive for as long as possible, just like dairy cows.

In a way, Algenol is akin to a dairy farmer milking a product that can almost be gulped down right from the cow's teat. "More than that," said Coleman, "we keep our cows hooked up to the milking machine all the time." The other companies in the market are more like cattle ranchers that fatten up their livestock for the slaughter, but with the added step that they still need to process and cook the "meat" before it reaches the customer. The numbers that Algenol gives are indeed impressive and equally unbelievable. Right out of the gate, it's aiming to achieve ethanol production rates of 56,000 liters per hectare annually. That's four times the estimated future performance for cellulosic ethanol and about 10 times what corn ethanol is expected to achieve over the coming decade. But Algenol doesn't plan to stop there. It's eyeing a production milestone of 93,500 liters down the road, more than double the volume of fatty oils expected to come from competing algae processes.[17] And remember, just as much fresh water is produced per hectare as ethanol.

But are we at risk of creating algae monsters here? There's a legitimate concern that algae that has been genetically engineered and modified in the laboratory could, when deployed on a large scale, eventually escape its confines and wreak havoc on natural ecosystems. "Use of genetically modified organisms on this scale is an untried experience for everyone and may not ultimately be accepted by the public," said R. Malcolm Brown Jr., a professor of botany and algae expert at the University of Texas.[18] Pictures and videos of oil-covered birds and dead fish highlight the dangers of oil exploration and development, but the environmental impacts of an algae project gone wrong could prove just as damaging. If a form of algae has been bred and genetically enhanced to survive under unnatural conditions, to function in essence as a kind of super algae, what happens if it seeps into the wild?

In Algenol's case, it's an unlikely scenario, the company says. The cyanobacteria are designed to turn all sugars into ethanol instead of carbohydrates. This modification effectively stunts the organism's growth, making it a weak specimen in the wild. It's like breeding a lapdog and then making it compete with a pack of wolves. As one biologist working on oil-producing algae said, "We're trying to make these guys couch potatoes. Big and fat and happy." There has even been talk of designing algae with "suicide genes" that would cause them to expire after a certain number of days in the wild. Still, public perception remains a powerful barrier to new innovation, and Algenol and others will have to satisfy lingering concerns. The speed at which algae can mutate and the associated risks should not be underestimated.

SCUM IN TRAINING

Paul Woods and marine biologist Harlan Miller took me for a walkthrough of Algenol's West Palm facility, beginning with the company's first-generation bioreactors, which look like oversized eavestroughs — white tubs about 19 inches wide and a foot high — lined up in tight rows and covered by a peak of hard, transparent plastic. True to their name, the blue-green algae have a bluish tint and could be seen floating within each bioreactor in a shallow layer of seawater. A hose, slightly thinner than the garden variety, was connected to each reactor and supplied the organisms with a steady meal of carbon dioxide gas. A thin wire stretching inside the length of the bioreactors is part of a mechanical mixing system that circulates the slimy content, making sure it gets maximum exposure to sunlight and to the CO_2 pumped into the enclosures. "This [setup] is older. We're still trying to develop a mixing system that is inexpensive to build and operate," Miller explained. Each bioreactor is also equipped with sensors that measure temperature, humidity, and CO_2 levels, which are remotely monitored through a wireless Ethernet network. Miller then drew my attention to the condensation, small droplets of water-diluted ethanol visible on the inside of the bioreactor cover. As the ethanol vapor escapes from the algae, these droplets form into bigger droplets and, thanks to gravity, eventually start traveling along the inside of the angled cover down the walls of the bioreactor. It's a continuous, low-cost way of collecting the ethanol prior to its purification. No external energy source is required.

We moved to another area of the outdoor facility to see a newer generation of bioreactors. These ones looked dramatically different. Think of a Freezie brand ice pop lying on its side but filled with air instead of flavored ice. Now imagine the soft, clear plastic packaging being nearly four feet wide and 50 feet long.

Just like in the older bioreactors, the reactor contains a layer of algae floating in shallow seawater, which is exposed to sunlight and CO_2 The organisms produce ethanol that evaporates inside the durable plastic body of the bioreactor and accumulates as condensation on the rounded reactor ceiling. And again, gravity carries the droplets of diluted ethanol down the sides of the plastic walls, where they are captured at the base in a simple trap that drains the fluid to a central collector.

To the average observer (me), the bioreactor appears like nothing more than a big plastic bag. Not quite, said Woods. "That freaking bag does it all." The plastic, he told me, is a proprietary material made by Dow Chemical that is used as a

A casual and cool Paul Woods (left) with Harlan "Lanny" Miller at Algenol's West Palm Beach facility in June 2010. Behind them sit a dozen or so experimental photobioreactors, or what Woods calls those "freaking bags" that do it all.

protective cover for photovoltaic cells. Unlike your standard-fare polyethylene, which gets beaten up by the outdoor elements and constant exposure to the sun, Dow's material is guaranteed for 35 years when applied to solar cells. For Algenol, this high-tech plastic represents about half the cost of its bioreactor, but the material's long-term durability is what makes the company's approach so feasible. "It's crucial," said Woods. "It completely changes the economics of how long these algae-ethanol plants last, even if you have to swap out the bag three times over 45 years." There's no point, he said, of building a massive algae-ethanol facility if you have to swap out bioreactors every few years. Longevity is crucial to the business model, even if there is a slight upfront premium.

At the end of my tour, we stopped into a makeshift field lab located beside the bioreactors, what one of Algenol's technicians called the "grow-op" room. It's basically a trailer loaded with shelves of bottled algae culture, each being fed CO_2 through an intravenous-like highway of plastic tubes. To simulate sunshine, shelves are backlit by a wall of fluorescent lighting. This is where the algae at this test facility begin their early growth after being "seeded" in a plastic Petri dish. From the dish, they go to a sterile flask, graduating to a large bottle as water and light levels are stepped up. The batches must be carefully acclimatized before transitioning to the outdoor bioreactors. "You have to be gentle with them at first," said Woods, picking up a bottle and pointing to the green goop inside. "This right here is the most sophisticated algae — ethanol-producing hybrid algae. This is the scale-up just before it goes to the bioreactor. This is the real deal." No wonder Woods got a tad angry when he realized the door to the trailer wasn't locked. "Your sole job today is to get a huge freaking bolt for this door," he said tersely to one of his staffers. "I don't want to see this door open like this again."

I'd be ticked off, too. An incredible amount of work goes into finding the right algae for the right job. The pool of specimens is worldwide, and getting to Algenol's Florida facility is like qualifying for the algae Olympics. Thousands of cyanobacteria are collected, analyzed, and put through a screening process at an Algenol biological laboratory in Germany that works in cooperation with the company's lab in Baltimore, Maryland, which claims to have the world's largest algae library. There are plenty of blue-green

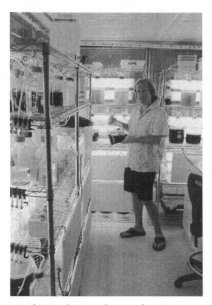

Paul Woods, standing in the "grow-op" room, shows off a bottle of his best ethanol-producing algae.

algae to sort through; scientists credit these single-celled creatures for shouldering up to 30 percent of the photosynthetic energy production on Earth.[19] Of the thousands of cyanobacteria that have been screened in Germany for their ability to produce ethanol, less than 100 make it to a lab in Spain that conducts outdoor testing. Some might look like champions under a microscope but that doesn't mean they can survive in a bioreactor as part of a massive commercial facility. It also doesn't guarantee that they can stay on their game after going through genetic enhancements. Of the group that make it to Spain, only the most elite algae have so far made it to West Palm Beach — less than 10 specimens. "It's a competition," said Woods. "Once they come here, we have to train them more. We have to train them and train ourselves to give them the best chance to produce [ethanol] over the long term."

ANGEL FROM MEXICO

The team-building doesn't stop with the algae. Woods knew back in 2005 that if he was going to pursue his algae-to-ethanol dream, he would need to assemble a management team that could carry the company through its difficult startup phase. The first recruit was Craig Smith, a veteran executive of the biosciences industry and founder of Guilford Pharmaceuticals, a maker of drugs for the treatment of brain cancer and other diseases. Guilford was sold for $178 million in 2005, and around the same time Woods attended a neighborhood party at Smith's house. "He didn't even show up for his own party," Woods recalled. "But I met his wife that night, and the wives and neighbors started talking about what I was doing." A week later, what had become a regular Thursday night party in the neighborhood moved to Woods' house. Smith and Woods finally met but only for five minutes. "He told me we'd talk in a week," said Woods, who at that point was wondering what this guy was all about. Was he trying to avoid Woods? "What I didn't understand, because I didn't know Craig, is that because he ran a biotech company he had unbelievable access to information. So when I met him a week later, he had already read our patents and had investigated a mechanical engineering paper we'd written. He literally walked into my house and said, 'I'm in.'" Smith is now Algenol's chief operating officer.

The third founding partner in Algenol, and the company's executive vice president, is Edward Legere. Another seasoned pharmaceutical executive, Legere is also an expert at helping companies scale up all the way from laboratory testing to pilot projects to full-blown commercial facilities — a mountain Algenol has yet to climb. But perhaps the most intriguing co-founder of Algenol is Alejandro González Cimadevilla, a Mexican businessman with family links to Grupo Modelo, the world's sixth-largest brewery known to most people as the maker of Corona beer. His father, Luis González Diez, is nephew to Grupo Modelo founder Don Pablo Diez and kept a senior role in the company until several decades ago when he founded Grupo Gondi, an industrial conglomerate with a focus on paper production, recycling, and packaging. Alejandro González, one of four brothers involved in the family business, is vice president of strategic planning at Gondi and sits on the company's board. The family still has strong financial ties to the brewery.

Woods wouldn't say exactly how he came to know González, a silent partner in Algenol who generally keeps a low profile as a director of the company. But he heaped praise on González for being an early financial backer and believing in the concept. "He and I put $70 million of our own money into this, but it was largely him," Woods said. "We have no venture capital money. If it hadn't been for him, as a really strong and faithful angel investor, Algenol simply wouldn't exist." In a way, González is to Woods what J.P. Morgan was to Nikola Tesla — a super angel who gave him the freedom to push forward on an idea that few others were willing to fund.

More than that, González became Algenol's first major customer. According to one report, González began to look seriously at renewable-energy investments after his wife got pregnant and, as most expecting fathers do, he questioned the state of the world he was leaving behind for his son. Details are sketchy, but it appears González founded a Mexico City–based company named Biofields in 2006 and began a search for investments in alternative energy and fuels. It was his financial broker who finally connected González with Woods, and the two men worked out a deal that would help fund development of Algenol while giving Biofields the license to develop a sprawling $850-million algae-to-ethanol project in the Sonoran Desert in northwest Mexico. That agreement, announced in 2008, would see Algenol bioreactors eventually blanketing more than 22,000 hectares of Mexican desert — about the area of 40,000 NFL football fields or a city like Boston. The aim is to produce nearly a billion liters of ethanol a year by 2014, ramping up to nearly four billion liters by 2020.[20] Sounds like a lot, but remember that the United States consumes 522 billion liters of gasoline a year, while U.S. ethanol mandates in 2010 targeted annual sales of 49 billion liters.[21] But it's a solid start. Construction on a small pilot plant reportedly began in 2010. Once Biofield's facility is in operation, CO_2 from a nearby power plant will be used to feed the algae. González plans to sell his company's ethanol to PEMEX, Mexico's state-run oil company, or to export the green fuel to neighboring California.

Woods was happy to ride the momentum that González brought to his company, but the wave got even larger in the summer of 2009 when Algenol announced it was partnering up with its bioreactor supplier, chemical titan Dow Chemical, to build a demonstration plant in Texas. Dow isn't in the fuel business, but it does rely heavily on fossil fuels in the production of ethylene, which is a basic chemical used for making many types of plastic. Like many Fortune 500 companies looking to clean up their corporate images, as well as reduce their reliance on petroleum, Dow saw Algenol's process as a way to lower its carbon footprint *and* have an on-site source of ethanol, which can be used instead of petroleum to make ethylene. "Our strategy is to put CO_2 to work," said Dow's director of biosciences Steve Tuttle, when the deal was announced. The original plan was to deploy 3,100 bioreactors on 24 acres of property at Dow's manufacturing site in Freeport, Texas. A nearby Dow chemical factory would have fed CO_2 to the bioreactors. Months later, Algenol got word that the U.S. Department of Energy was tossing a $25-million grant at the project. But Dow, according to Woods, had trouble convincing regulatory authorities to approve use of the land because of the project's potential impact on nearby wetlands, so in late 2010 the companies decided to move construction from Texas to 40 acres of land next to Algenol's new headquarters in Lee County, Florida. Not quite what was originally envisioned, but Woods called it a more "expeditious route" that would result in a slightly larger facility being built by the end of 2011.

"Our business really changed a lot since we got Dow as our primary partner," Woods told me as we drove to a nearby Tex-Mex restaurant for some nachos and beer. Shortly after the Dow deal was announced, Algenol struck a partnership with the Linde Group, the top supplier in the world of industrial processes for handling gas. Linde is a pro at piping carbon dioxide to and from industrial facilities. For example, it supplies CO_2 through pipelines to more than 500 greenhouses that together cover about 1,500 hectares. Algenol and Linde are now working together on a similar system to transport CO_2-rich emissions from power plants and other industrial facilities to Algenol's bioreactors. Another feather in Woods' cap came in May 2010, when Valero, the largest independent petroleum refiner and marketer in North America, revealed a development agreement with Algenol. Valero is also the third-largest ethanol producer in the United States, so it has an interest in any next-generation ethanol technologies that are more sustainable and have a chance of working on a large scale. Valero, mind you, has invested in several biofuel startups, including algae-oil company Solix Biofuels, but its link-up with Algenol was further proof that the big, high-profile boys of U.S. industry are taking notice and getting involved.

Having Dow, Linde, and Valero on board put Algenol in a different league, even if it would still remain in training camp for a while. Woods was clearly buoyed by the alliances he had built. "We've got the CO_2 side covered, we've got the chemical side covered, and lastly the fuel side of the business covered," he said. "We have the upstream and the downstream." When I spoke to Woods again in November 2010, he was just about to hop on a plane to meet with one of Europe's largest energy companies, Spain's Abengoa Bioenergy, which ended up signing on as a partner. He also met with Saudi Arabia's state-owned Saudi Aramco, the biggest oil company in the world. "They're slow to move, but I've been at it for a year," he said, explaining that Valero was managing the relationship with the Saudis. "Behind the scenes, there's a lot of working going on with those guys."

But there's also a lot of skepticism building. "Algenol is a company that has hitherto overpromised and under-delivered," wrote Joshua Kagan, an energy analyst with GTM Research in an article that called into question Valero's interest in Algenol. He took issue with Paul Woods' "comical" claim in 2009 that Algenol would have 500 million gallons of ethanol in production by 2011. It was a fair comment. The company, as of summer 2011 and despite the promising early announcement from Biofields, hasn't come close. The media and industry analysts don't take kindly to boasts and exaggeration, and in this case Woods may have broken the cardinal rule, as many other algae biofuel companies have and continue to do. But Kagan kept an open mind, partly because of the blue chip partnerships that Woods has been able to assemble. These relationships, wrote Kagan, are hard to ignore — even for the most jaded. "Some might be skeptical and believe that if oil companies invest in advanced biofuels, they are doing so only to own the most promising technologies so they can later shelve them and keep consumers dependent on oil. Another perspective is that oil companies are in the 'liquid fuels' business, and as long as humans use liquids rather than electrons to fill their tanks, the oil companies have a vested interest in controlling that liquid, whether it is oil, ethanol, or designer gasoline."[22]

The question is whether the major oil companies, the ones that are serious, are serious enough. Take the BP oil spill, a single event with cleanup costs expected to exceed $40 billion, and that's excluding those costs that are more difficult to calculate, such as impact on local hotels, boat charters, restaurants, and other businesses hit hardest by the spill, not to mention the long-term economic impact on gulf fisheries and tourism.[23] A single event at $40 *billion*. If companies such as BP are prepared to swallow that kind of risk on oil, why don't they take the same risk on clean, more sustainable sources?

TIPPING THE SCALES

The answer may have something to do with scale. If you thought at this point Woods had defeated the odds and jumped most of the big hurdles necessary to disrupt the energy world, think again. The hardcore R&D may be over, the patents filed, the partners lined up, and deployment and demonstration underway. Assuming there are no fatal glitches along the way, the next mammoth challenge for Woods and Algenol's partners will be commercialization — heavy on capital, light on odds. "Scale is everything in this industry," said Coleman. Venture capitalist and biofuel investor Vinod Khosla likes to ask himself this question before throwing money into any energy company: "Can it produce, store, and disseminate on a wide enough scale to be material in carbon emissions reduction?" Even better, can it pass the "Chindia test" — can it be embraced in China and India, the fastest-growing markets and carbon emitters in the world? "If it doesn't scale, it doesn't matter," Khosla has said.[24] It's not that he doesn't believe algae-based fuels hold potential as a solution to climate change and the "war on oil." It's just that an algae company has yet to convince him it can compete with oil on cost and scale.[25]

Coleman told me that a key milestone for Algenol will be proving that its genetically modified cyanobacteria and its latest bioreactors, which are designed to hold 4,500 liters, can operate flawlessly under the same conditions as a full-scale commercial deployment. After that, said Coleman, building larger facilities — such as what Biofields envisions — will be like assembling Lego blocks. If one works, 100 will work, and if 100 work, there's no reason to believe 1,000 won't work — and up from there. That milestone is likely to be reached first in Lee County, Florida, the home of a 40,000-square-foot R&D facility that Algenol opened in fall 2010, and where the West Palm facility that I visited has since moved. This is where Woods and Co. will put their pre-commercial bioreactors through a barrage of real-world tests. The site, including the relocated pilot site being developed with Dow, will be watched closely to see if Algenol has what it takes to go big. I posed the scale question to Woods, telling him that some people in the industry say he'll never be able to do this on a scale that matters. "That's a lie. That's an absolute lie," he snapped back, insisting that getting to the scale that Biofields envisions can be done in three or four years. "But if you do nothing, you're never going to accomplish it in three years. Then it's just a self-fulfilling prophecy."

In its quest to go big, however, Algenol will also have to prove that it's better than the other little guys out there with similar ambitions — and not just those trying to squeeze oil from algae. In a world where "carbon" is increasingly a dirty word, and where a tax or some other price on carbon is considered inevitable, there are two options: catch and store the carbon, or catch and convert that carbon into some other useful product. Algenol says every 125 gallons of ethanol it makes consumes a ton of CO_2, but it and its algae rivals aren't the only ones trying to turn sunlight and CO_2 into fuel.

In 2009, I visited Sandia National Laboratories and had a

chance to watch scientist James Miller demonstrate his awkwardly named counter-rotating-ring receiver reactor recuperator, or the CR5, which uses the intense heat of concentrated sunlight to bust up the chemical bonds in water and CO_2 so their molecules can be reassembled into a synthetic gas and refined into gasoline or jet fuel. It looks like alchemy, but Sandia's "sunshine to petrol" initiative is the real deal. "What we're doing is analogous to photosynthesis," said Miller. "Algae are a good step in the right direction, but what we're trying to do is use more sunlight at higher efficiency." The potential for some of these unconventional approaches is so great that the U.S. Department of Energy said in 2010 that it would award up to $122 million over five years to a team of scientists at the California Institute of Technology and Lawrence Berkeley National Laboratory that is trying to develop fuels from sunlight and CO_2. The government's aim is to support the creation of a "direct solar fuels industry" that could help wean the country from oil.[26]

Making fuel isn't the only avenue. A Silicon Valley company called Calera, a potential game-changer in its own right, says it can capture a ton of CO_2 for every two tons of cement it produces. It does this the same way corals and shellfish pull calcium and carbon out of seawater to make their skeletons and shells. Calera has a pilot cement plant in California that is absorbing 86 percent of the CO_2 from the flue gas of a nearby natural gas–fired power plant.[27] Renewable fuel or green cement? We could use lots of both.

How about baking soda? A Texas company called Skyonic has figured out how to capture CO_2 (and other pollutants) from industrial plant emissions to produce bicarbonate, which can be used to make baking soda, detergents, and all sorts of industrial products. If carbon prices are high enough, it may also be worthwhile to just shovel the bicarbonate into a big hole in the ground, keeping it permanently buried and out of the carbon cycle.

The power industry's chosen path is less inspiring and more predictable: huge volumes of CO_2 from coal plant emissions are captured and pumped underground for permanent storage, either in depleted oil and gas wells or deep saline aquifers. The idea with so-called carbon capture and sequestration (CCS) is to treat CO_2 as dirt that should be swept under the rug — or as I like to say, to hide our peas in our potatoes. Let's ignore the fact that it's costly to do it and would require the development of a global network of CO_2 pipeline and storage infrastructure on par with what we already have for natural gas and oil. One problem with CCS is that it is energy intensive, meaning a power plant has to burn more fuel — and emit more CO_2 — to run the equipment and chemical processes required to capture those same emissions, concentrate them, compress them, pump them through a pipeline, and inject them underground. (Algenol, by comparison, uses free sunlight to power its CO_2 capture and it doesn't have to concentrate its CO_2 or compress it. Everything in its process is done under relatively low pressure, and gravity drives its ethanol collection.) Then, once the CO_2 is forced underground, we cross our fingers and hope nothing leaks into the surrounding environment, where it can contaminate drinking water or further acidify the ocean. Sudden leaks on a large enough scale, possibly triggered by an earthquake, could potentially suffocate animals (including humans) in an affected area. Small but continuous leaks would also undermine the sole purpose of CCS, and there appears to be no way to indefinitely guarantee such leaks won't occur, or that they could even be detected. The risks, in the form of liability alone, are enormous.

"Good luck." That's what Microsoft founder and billionaire Bill Gates, now an active investor in energy technologies, once said when talking about the future prospect of CCS, a technology that he actually supports. "No private enterprise could take on the risk, and no government has stepped up."[28] Yet this is the approach the power industry continues to embrace. Costs vary for the technology, but estimates range from $80 to $120 for every ton of CO_2 captured and stored underground. Ask Paul Woods about CCS, and he can't help but rip the idea apart. "You're basically just taking $100 bills and shoving them in a hole in the ground," he told me. "If you give me $20 or $40, I'll take your CO_2 and give you a product for it." But that kind of reasoning doesn't yet resonate with the industry. Again, just as BP is willing to push the limits of drilling technology, go to more dangerous lengths to find and pump oil, and risk having to pay billions of dollars in damage-control costs in its continuing hunt for an increasingly hard-to-find resource, the world's power plant operators are willing to take the risks that come with blanketing the planet with CO_2 pipelines and pumping the world's most notorious greenhouse gas out of sight, but not necessarily out of the equation. For them it's about survival, about protecting an infrastructure that has kept them rich and has grown so large that changing direction, at least in any meaningful way, is not an option. Simply put, they have scale, they have control, and they don't want to lose it. The irony is thick, as these are the same companies that dismiss alternative energy technologies as too risky, prohibitively expensive, and impossible to scale, as if what they themselves are pursuing is a walk in the park. What's possible or not is clearly in the eye of the beholder.

SCRAPPY UNDERDOG

Still, there have been some eye openers. On July 14, 2009, the world's largest publicly traded oil company, ExxonMobil, said it planned to invest $600 million over several years in algae biofuel research and development. Of that, it committed roughly half to genome pioneer Craig Venter and the company he founded in 2005, Synthetic Genomics, which is genetically engineering photosynthetic algae that can secrete their oils, similar to how Algenol's cyanobacteria naturally produce and emit ethanol. Venter has delivered before. He and his privately funded team of researchers were the first to sequence the entire human genome, a historic scientific feat they announced in 2000. They did it in less than a year and three years ahead of the public Human Genome Project. Keith Johnson, an environmental reporter with the *Wall Street Journal*, wrote that the partnership between Exxon and Venter "could mark a coming age of alternative fuels."[29] Indeed, only a few years earlier Exxon was one of the most vocal skeptics of human-caused climate change. Suddenly, it was leading the charge and putting algae on the front lines.[30] If concern over climate change wasn't a motivation, then certainly the coming impact of peak oil was forcing Exxon to pursue renewable alternatives.

The Exxon-Venter alliance has raised the bar for algae ventures focused on oil. No longer is it enough to just extract oil from algae; the approach du jour is to get the algae to hand it over on its own, cutting out the need to harvest, kill, grind, and squeeze. The new buzz term is to produce oil as a "drop-in solution" — a product that integrates seamlessly with existing oil infrastructure without the need for pre-treatment or refining (what I called a "shoehorn solution" in chapter 4). Already, there are a handful of other ventures pursuing this Holy Grail of biofuels, including Joule Unlimited, a low-profile company from Massachusetts, which has genetically engineered cyanobacteria that reportedly secretes both oil *and* ethanol. The competition is good and will force Algenol to up its game. Already, Woods is exploring ways to produce isobutanol (isobutyl alcohol), which packs a greater energy punch than ethanol, can be blended with gasoline, and is a building-block chemical for the production of jet and diesel fuels. He said his cyanobacteria can certainly produce the fuel. "We're just trying to figure out how to economically get the isobutanol out of the culture."

I asked him about the Exxon-Venter deal, and he downplayed the relationship. Woods has a sense of what's going on at Exxon, as several of his own employees were recruited from the oil giant. "The company talks a game about CO_2 mitigation, but if you look at the $300 million they announced for Synthetic Genomics, the actual funding was below $20 million." Exxon, in fact, did make clear certain strict milestones must be met before the next round of cash is released. "The problem with that research program is it's designed to take a long time," said Woods. "If you want something done more quickly and effective like Dow, Linde, and Valero do, then you go a different route; you go with Algenol."

Bold talk from a man on the front lines who must know better than anyone else that few will be left standing after the field is rushed. It's the nature of the game. Woods is an underdog, a scrappy one at that, but you have to be scrappy to stay alive for the next fight and the one after that. He's also in a rush to get out of the lab, to escape the endless calls for more R&D, to get out there and actually do something that can start having an impact. And he challenges anyone who has doubts about Algenol's technology and approach to do their due diligence, just as Dow, Linde, and Valero have. "When you examine technology very carefully, you can easily whittle down the bullshit from what's real. If you really want to examine things, to separate the wheat from the chaff, you can." By 2020, perhaps earlier, we'll be able to look back at Algenol and determine whether it has been a success or failure. Either way, it's a spectacular attempt that's contributing to the advancement of a more sustainable world. "If you do nothing, of course you're not going to get there," said Woods.

Algenol has proven adept at building friendly alliances, something Dick Weir and his mysterious company EEStor is less accomplished at. Even so, as you'll find out in the next chapter, Weir has both believers and skeptics watching his company's every move. EEStor says it has developed an energy-storage device that could end the reign of gas-powered vehicles and usher in the dominance of renewable energy. But as astrophysicist Carl Sagan once said, "Extraordinary claims require extraordinary evidence."

Notes:

1 Cyanobacteria are not algae in the narrow sense: they are non-plant microorganisms capable, like true algae, of photosynthesizing sunlight. But they are regularly thrown under the algae umbrella. For this reason, and for the purposes of this chapter, cyanobacteria and algae will be used interchangeably.

2 For the remainder of this chapter, all uses of "algae" refer to microalgae, as opposed to macroalgae (which is seaweed).

3 The U.S. Environmental Protection Agency ruled in October 2010 that newer vehicles can operate on a maximum 15 percent mixture of ethanol in gasoline. Prior to that, the blend limit was 10 percent. To get up to 85 percent a "flex-fuel" vehicle or retrofit is required, but this now costs less than $300 per vehicle.

4 Andrew Lacis, "CO$_2$: The Thermostat That Controls Earth's Temperature," Goddard Institute for Space Studies website. October 2010. http://www.giss.nasa.gov.

5 William Neuman, "Rising Corn Prices Bring Fears of an Upswing in Food Costs," New York Times. October 12, 2010.

6 M.R. Schmer et al., "Net Energy of Cellulosic Ethanol From Switchgrass," Proceedings of the National Academy of Sciences of the USA, University of Nebraska. Volume 105, no. 2. (January 15, 2008): 464–469.

7 Tom Nugent, "Energy to Burn," Western News website. November 17, 2009. (This profile of Paul Woods appeared in the University of Western Ontario's online news service.)

8 Jean Gruss, "Fuel the World," Gulf Coast Business Review. December 11, 2008.

9 Anastasios Melis et al., "Trails of Green Algae Hydrogen Research — From Hans Gaffron to New Frontiers," Discoveries in Photosynthesis (University of California, Berkeley). Volume 20, IX (2005): 681–689.

10 T.J. Lundquist et al., "A Realistic Technology and Engineering Assessment of Algae Biofuel Production," joint academic study published by the Energy Biosciences Institute, University of California, Berkeley. October 2010.

11 H.C. Greenwell et al., "Placing Microalgae on the Biofuels Priority List: A Review of the Technological Changes," Journal of the Royal Society. 7, no. 46 (May 6, 2010): 703–726.

12 John Sheehan et al., "A Look Back at the U.S. Department of Energy's Aquatic Species Program: Biodiesel From Algae," National Renewable Energy Laboratory. July 1998.

13 T.J. Lundquist et al.

14 Emily Waltz, "Algae Energy Orgy," Mother Jones. September/October 2009.

15 John Sheehan, "Engineering Direct Conversion of CO$_2$ to Biofuel," Nature Biotechnology, no. 27 (2009): 1128–1129.

16 Mark S. Wigmosta, Andre M. Coleman, Richard J. Skaggs, Michael H. Huesemann, Leonard J. Lane. "National Microalgae Biofuel Production Potential and Resource Demand." Water Resources Research. April 13, 2011. http://www.agu.org/journals/wr/wr1104/2010WR009966/.

17 John Sheehan.

18 Email to author from R. Malcolm Brown Jr. July 14, 2009.

19 John M. Pisciotta et al., "Light Dependent Electrogenic Activity of Cyanobacteria," PLos One (Public Library of Science). May 25, 2010.

20 Veronica Diaz Favela, "Big Plans for Ethanol From Algae," Inter-Press Service/IFEJ. December 4, 2009.

21 Biofields has lowered expectations since its original 2008 announcement. Originally, it planned to produced 7.6 billion liters by 2020, a figure it has since cut in half. Also, construction of its pilot plant and plans for its first commercial plant have also been pushed back, reflecting the difficulty of getting new technology out of the lab and into the field.

22 Joshua Kagan, "Valero Invests in Algenol: What's Going On?" Greentech Media. May 10, 2010.

23 Graeme Wearden, "BP Oil Spill Costs to Hit $40 bn," Guardian.co.uk. November 2, 2010.

24 Martin LaMonica, "Khosla: Crazy Clean-Tech Ideas Yield Breakthroughs," CNET News. September 24, 2008.

25 Vinod Khosla put out a draft of his "green investing strategies" in 2007 in which he emphasized the need for any clean technology solution to scale if it is to have any meaningful impact.

26 Josie Garthwaite, "DOE Awards Up to $122M for Making Fuel from Sunlight," Greentech Media. July 22, 2010.

27 Claire Cain Miller, "Mixing in Some Carbon," *New York Times*. March 21, 2010.

28 Tyler Hamilton, "Overhauling Energy Will Be Slow, and Expensive," *Toronto Star*. August 16, 2010.

29 Keith Johnson, "Biofuels Bonanza: Exxon, Venter Team Up on Algae," *Wall Street Journal* Environmental Capital blog. July 14, 2009. http://blogs.wsj.com/environmentalcapital/2009/07/14/biofuels-bonanza-exxon-venter-to-team-up-on-algae/.

30 Jeffrey Ball, "Exxon Softens Its Climate Change Stance," *Wall Street Journal*. January 11, 2007.

CHAPTER 6

Secrecy in Cedar Park

Raising the Bar on Energy Storage

"A great idea isn't the next guy doing a lithium-ion battery, almost certainly."
— Venture capitalist Vinod Khosla

"EEStor disrupts so many different sectors — it's the end of fossil fuels."
— Ian Clifford, founder, ZENN Motor Co.

Former IBM engineer Richard Dean Weir doesn't want attention. "I don't want to be a celebrity," he told me. "I really don't need that." He generally loathes the media. His company EEStor, founded in 2001, doesn't have a corporate website (at the time of writing), and it rarely puts out a press release. What words Weir does use are few and abrupt. I first reached out to him in February 2006 to research a newspaper feature; the email he sent back was dismissive and terse. "EEStor would like to have you and your paper not publish any articles about our company." He

175

went further, telling me I wasn't authorized to publish his dismissive response. That was a career first. But when someone tries to play hard to get, people naturally try to get him. Despite Weir's desire to stay out of the spotlight, EEStor has developed a cult-like following on the Internet, where the rumor mill swirls and skeptics and believers trade barbs as they wait for Weir to deliver on a promise that's supposed to change the world as we know it. Those who track and document the company even call themselves EEStorians.

So what explains this unusual fascination normally reserved for the latest i-gadgets from Apple or new video game releases? In two words: energy storage, a technology we take for granted and on which we increasingly depend. What Richard "Dick" Weir is working on, however, is not your run-of-the-mill Duracell battery, or even the more advanced lithium-ion batteries that are powering most of the plug-in hybrid and all-electric vehicles coming soon to a dealership near you. EEStor says it has developed a new type of energy storage device, one based on research from the Reagan-era "Star Wars" program, which could usher in a future of low-cost electric cars capable of traveling more than 400 kilometers on a single five-minute charge. That alone is jaw-dropping, but Weir also insists his electrical energy storage unit, or EESU, would last longer than the life of any vehicle it powered. For that matter, it would outlast any laptop, smart phone, e-reader, electric scooter, cordless tool — any portable consumer product that runs on electricity or, as you'll see, any military weapon. Imagine having an iPhone that only needs charging once every two or three months. According to one of EEStor's patents, this remarkable storage device would have the added benefit of not being "explosive, corrosive, or hazardous," which can't be said for most other battery chemistries. "You could drive a metal stake through it," Weir said in one of his rare interviews. "It will be the safest battery the world has ever seen."

But just as disruptive would be its application to renewable energy, such as wind and solar, which has historically been limited because of its inability to generate power when factories, businesses, and homeowners always need it. There exist today some forms of large-scale energy storage, such as pumped storage of water or compressed-air storage, but these are expensive and only possible in certain locations and geographies. EEStor's technology would bust open the market by offering a way, not limited by location, to affordably store huge quantities of electricity from wind turbines that don't spin all the time or solar panels that are at the mercy of thick clouds and setting suns. Massive storage facilities using EEStor technology could then dispatch that inexhaustible, emission-free power the same way we do now with coal and natural gas power plants. Low-cost energy storage along the lines of what EEStor envisions could level the playing field, once and for all, between renewables and fossil fuels, dramatically changing the way we design electricity systems and how we power transportation. It would be a true breakthrough, that elusive "game-changing" technology that whips up both optimism and skepticism. Such a form of cheap energy storage would permit us to power more of the planet with clean, renewable energy — instead of coal and natural gas — that in turn would be used to charge up electric cars that have range and performance to rival vehicles fueled by gasoline and diesel, which would eventually become dinosaurs of a dirty era. The result is a virtuous cycle of clean energy that makes it possible for world economies to phase out, over time, the heat-trapping, air-polluting fossil fuels they have come to depend on. "Honestly, I don't know if EEStor has solved it or not," said Elon Musk, co-founder and chief executive of electric-car maverick Tesla Motors. "I've heard people say it's just bullshit, and others say it's a big breakthrough. Until you see something on the road, objectively, it's hard to say what's true." So we impatiently wait, some more doubtful than others but all, whether they admit it or not, carrying a seed of hope.

EEStor first landed on my radar in late 2005 when I stumbled upon a short news story published by *Businessweek*. It mentioned that the Texas-based company had raised $3 million as part of a small financing deal led by famed venture-capital firm Kleiner Perkins Caufield & Byers. EEStor, according to the article, claimed to make a battery "at half the cost per kilowatt-hour and one-tenth the weight of lead-acid batteries."[1] Not much information to go on at the time, but enough to get the imagination charged up. General Motors' short-lived EV1 electric car, produced during the late 1990s and more recently profiled in the critically acclaimed documentary *Who Killed the Electric Car?*, ran on 26 lead-acid batteries that together weighed about 1,300 pounds, or more than 40 percent of the vehicle's weight. At best, it could travel 100 miles on a single charge. EEStor's energy-storage technology, by comparison, would cost half as much and could technically, on a weight-for-weight basis, give the EV1 a single-charge range of 1,000 miles.[2] One could easily dismiss such a claim as wishful thinking, but it was the connection with Kleiner that leant it credence. The venture capital firm was legendary in Silicon Valley circles, having made early and stunningly successful bets on Google, Amazon, and others that are now giants of the Internet ecosystem. From the outside looking in, Kleiner's involvement with EEStor was a credible and curious endorsement.

The timing was also ideal. In 2005, the public had grown tired of the never-ending promises of hydrogen-powered fuel cell vehicles. Electric cars — from Toyota's hybrid-electric Prius to Tesla Motors' planned all-electric Roadster — were beginning to capture more headlines and investment dollars. Energy security was top of mind in a post-9/11 world and climate change was set to become a mainstream concern in 2006 with Al Gore's *An Inconvenient Truth*. Meanwhile, organizations such as the California Cars Initiative (CalCars) saw the re-emergence of the electric car as inevitable and pushed hard to have major automakers begin development of plug-in models that ran mostly (or exclusively) on juice from a wall socket. The established automakers resisted at first, but most of them eventually fell in line, putting unprecedented focus on the need to develop better battery technologies. Mysterious EEStor, whether one took it seriously or not, stood out by every measure.

THE PRE-EESTORY

So who is this Dick Weir fellow, anyway? Details of his early professional life are sketchy, feeding the mystery and skepticism around EEStor. Those who know Weir say he grew up in Montana and as a young adult enroled with the U.S. Marine Corps, which had him flying fighter jets off aircraft carriers in northern Japan during the early 1960s, part of routine patrols that kept an eye on Soviet naval activity around the port of Vladivostok. Later that decade, Weir got a degree in electrical engineering from California State Polytechnic College in Pomona. During the 1970s and 1980s, he worked for a variety of companies in computing, including IBM, Burroughs, Xerox-PARC, and Micropolis. At these companies, he developed an expertise in disk-drive technology and integrated circuit fabrication, while also learning the importance of materials science in their advancement. Weir shifted gears somewhat in the 1980s after President Ronald Reagan launched his Strategic Defense Initiative ("Star Wars"). At some point, Weir began working in the space and technology division of military contractor TRW, which was developing laser-based defense systems for the Star Wars program. The exact nature of this work is unclear, but in a confidential investor interview that leaked onto the Internet in July 2009, Weir, talking about the origins of EEStor's technology, said he "invented a chip that went into deep space" and, under contract with the Central Intelligence Agency, invented an ultra high-frequency channel.[3] "I'll be very blunt," one business source close to Weir later told me. "I think Dick was working on a deep-space channel that could control satellites that were designed to shoot down missiles, but it needed to be powered." That power, in deep space, would have to come from an extremely advanced energy storage device.

It was during this time that Weir met Carl Nelson, the other, even more enigmatic founder of EEStor. Nelson was educated and worked for many years at the Massachusetts Institute of Technology, where he studied under Arthur von Hippel, a pioneer in molecular engineering and a world-renowned expert in advanced materials. Nelson himself became an expert in crystallography (the study of how atoms are arranged into solids) and the development of ceramic powders that can be used in a range of high-tech products. These included "dielectric" materials that control the flow of electrons in an electric field. Nelson was a smart cookie, and he began working in close collaboration with Dick Weir at TRW.

Shortly after the end of the Cold War, and under the order of incoming President Bill Clinton, the space-based Star Wars program envisioned by Reagan — and long protested by the Soviet Union — began to shift focus to a ground-based missile defense system that was more regional in dimension and less threatening in scope. Military funding for the type of expensive work Weir and Nelson were doing began to dry up, and they never got around to completing the power storage part of their work. "The Russians threw in the towel and we [the U.S.] didn't need it any longer," Weir said in the leaked interview. "So I shelved it and went into doing some work for the disk-drive industry." Weir and Nelson knew the path they needed to pursue on energy storage, but getting on track without military support proved challenging. Some of the high-voltage semiconductor circuitry needed to build their envisioned device, for example, was prohibitively expensive at the time.

The computing revolution was well underway, and the Internet, including this new invention called the World Wide Web, was just beginning to grip the public's imagination. So Weir and Nelson shifted their focus to what they both knew well. They founded Tulip Memory Systems, a disk-drive storage company in Fremont, California, and through the early 1990s began filing patents on a "groundbreaking" magnetic disk-drive storage technology they had developed. They came up with a way to manufacture a memory-storage disk with a thin-film surface of titanium alloys, one that would be capable of storing far more data than industry-standard aluminum disks at the time. Micropolis, one of Weir's former employers, helped fund their first few years of research and development.

In 1996, however, Tulip began to wilt. Micropolis ran into its own financial troubles, and by the end of 1996 the funding lifeline for Tulip was cut off. It was around then that Weir and Nelson met Gary Hultquist, a California investor and entrepreneur who was intrigued by their work. Hultquist led a group of angel investors who breathed new life into Tulip with a $2.25 million investment. He also arranged $5 million in additional investment from Titanium Metals, the world's largest supplier of titanium. As part of the deal, they changed Tulip's name to the more high-tech-sounding Titanium X. "I thought at this point we had the world by the tail," recalled Hultquist, explaining that disk-drive giant Seagate Technology had signaled that they would embrace Titanium X's technology if strict specifications could be met. "We kept improving it each year, and we finally did meet the standards given to us by Seagate." But it took longer than expected, and Seagate had a change of heart during a pivotal meeting in 2000. "They told us they wouldn't end up switching to titanium disks because switching costs through the supply chain would be too great, and titanium prices were too volatile." Weir, Nelson, and Hultquist were stunned that Seagate had shut the door and, as a result of this unexpected blow, the decision was made to shutter Titanium's business. "When I left them, they had no money and our project had been a failure," said Hultquist, now a principal at investment firm NewCap Partners in San Francisco.

About six months later, Weir and Nelson began filing energy-storage patents under the name EEStor. I asked Hultquist what the two men were like, both personally and professionally. "There was a lot of great engineering done by Dick and Carl, and while they missed several deadlines they eventually performed," he told me. He also painted a picture of Weir that was in stark contrast to Nelson. Both men were in their late 50s or early 60s at the time. Weir, a balding, paunchy fellow about six feet tall, was an abrupt, dogmatic, and closed individual who didn't work well with outsiders but was full of energy. "He looks people directly in the eyes," said Hultquist. "He always told me without question we would get things done. You can suggest things to him, but you can't tell him to do something. Dick does things his way." Nelson, on the other hand, was shy and spoke mostly through Weir. He was a tall, gaunt, and sickly looking chap, described by one source as a "scary professor" who did little else but work. "I had an office about 20 or 30 feet from Carl," said Hultquist. "I would see him all the time, but he was always at his desk." In a way, Weir was the yin to Nelson's yang.

BATTERY BOTTLENECK

What Weir and Nelson described in their first round of EEStor patents wasn't a new type of battery per se. In fact, they claimed it as a "replacement of electrochemical batteries" that are in popular use today. A battery is basically a device that converts chemical energy into electrical energy. Take two pieces of metal — zinc and copper, for instance — and separate them by a mixture of salt and water. This "salt bridge" in the middle is called the electrolyte; the zinc is a positive electrode and the copper a negative electrode. The electrolyte contains what are called free-flowing ions, and when you attach a load, such as a flashlight bulb, to the end of each electrode, the ions in the electrolyte will flow from the zinc to the copper. That flow of ions produces a direct current of electricity. This is basically what Italian physicist Alessandro Volta demonstrated for the first time back in 1800, and while there are many flavors of battery chemistries today, the basic principle — that of an electrochemical reaction between two electrodes — still holds. For example, lead-acid batteries first invented in 1869 still use lead-based electrodes separated by an electrolyte of sulphuric acid; newer lithium-ion batteries use lithium-based electrodes separated by an electrolyte made up of lithium salts.

Batteries are everywhere, from the throwaway "primary" batteries used in flashlights and wristwatches to the rechargeable "secondary" batteries used in traditional cars (lead-acid), hybrid-electric vehicles such as the Toyota Prius (nickel-metal hydride), and laptop computers and smart phones (lithium-ion). They can be designed to store small or large amounts of energy, but they all release that energy in a slow, steady current. In many instances, the limitation of battery technology at a given point in time has been a bottleneck to other innovations. Take smart phones. Each new added feature — color screen, touch screen, digital camera, MP3 player, Wi-Fi, and so on — consumes energy. At the same time, consumers demand smaller gadgets that can operate longer between charges. Phone makers can make their device more energy-efficient, but there is also immense pressure on battery makers to develop smaller, lighter, and cheaper batteries that hold more energy. It's been a tall order for companies such as Apple or Research In Motion, as they try to keep one step ahead of each other in the smart phone wars.

The same battery bottleneck has held back development of electric cars. The challenge has been to find a battery chemistry that can take up the least amount of space, weigh as little as possible, power a car for several hundred kilometers, *and* be affordable enough for an electric vehicle to be competitively priced with its gas-powered equivalent. One key metric is specific energy, which is the amount of energy (watt-hours) a battery can supply per kilogram of its weight. Lithium-ion batteries, with a specific energy of between 80 to 150 watt-hours per kilogram, have so far been anointed the battery chemistry of choice for electric cars. By comparison, the best lead-acid batteries have a specific energy of only 32 watt-hours per kilogram. Still, despite their superior performance, the best lithium-ion batteries today still fall well short of where we'd like to be. "A number of incremental improvements are underway, but they will at best offer two-times improvement in price performance," wrote venture capitalist Vinod Khosla in one of his many online missives on new energy technologies. "They represent improvements, but not radical changes, to techniques used in batteries for consumer electronics. We cannot expect significant increases in performance absent fundamentally new approaches."[4]

Actually, it wouldn't be so bad if you could recharge a lithium-ion battery in the same time it would take to fill up your car with gasoline. But we can't, at least not yet. There are expensive quick-charge technologies out there, but even they still require 15 to 30 minutes to perform a decent recharge. Modern consumers are creatures that demand instant gratification. We live in a culture of impatience. We hate to wait. We want it now. We'll do it if we have to, but most of us would rather not have to wait several hours to recharge a device or vehicle. Another strike against batteries is that they tend to be made of toxic, corrosive, flammable, and/or explosive materials that are difficult to recycle and risk leaving us a shameful landfill legacy. They also degrade over time with each recharge cycle, meaning they often expire well before the products they are designed to power.

There's another type of technology that gets around many of these limitations: the ultracapacitor. An "ultracap" is basically two metal plates of opposite charge that are separated by an insulator, or what's also referred to as a dielectric. Think of a peanut butter sandwich, where the two pieces of bread are the metal plates and the peanut butter is the insulator. If you apply voltage to the sandwich, the peanut butter prevents the current from flowing through, so instead electrons accumulate on the bread and create an electric field across the peanut butter. The Earth, in a way, is part of one big spherical capacitor. The ionosphere is one plate and the surface of the Earth is the other, with more than a hundred kilometers of lower atmosphere in between acting as the dielectric. Solar wind from the sun constantly fills the ionosphere with positively charged particles, and when that charge builds up too much, the electric field in the atmosphere reaches a breaking point and — *bang!* — lighting strikes. William Tahil, research director with Meridian International Research, has been exploring ways of economically tapping into this "atmospheric electricity" as a new source of clean energy since 1997.[5]

The electric field that builds up inside an ultracapacitor is also similar to the electric field that builds up on your body (one plate) when you walk across a dry carpet in socks. And, similar to how one can release a sudden shock when reaching through the air (the dielectric) and touching a doorknob (the second plate), so too can an ultracapacitor dispatch a powerful burst of energy. No chemical reactions take place. No toxic or corrosive materials are used. Ultracapacitors, unlike batteries, can also be instantly charged, discharged, and recharged almost infinitely. The problem is that ultracapacitors don't store much energy. They pack a powerful one-time punch, but on a watt-hour per kilogram basis, lithium-ion batteries hold 100 times more specific energy.

To get around the limitations of batteries and ultracapacitors, Weir and Nelson say they have developed a "ceramic battery" that is a hybrid of a battery and ultracapacitor, combining and enhancing the best features of both. At its core is a cell of two aluminum plates that sandwich a dielectric of "composition-modified barium titanate," which is a fancy way of saying a white ceramic powder with an amazing ability to hold an electric field. Each cell is about the size of a microchip and one-fifth the thickness of a human hair. One hundred cells are packed tightly into an element, and 10 elements are tightly packed to form a component. An EEStor ceramic battery that could power an electric car would contain more than 30,000 components or more than 30 million cells. Pound for pound, it would store twice the energy of the lithium-ion technology used in the Tesla Roadster, but Weir and Nelson have said they could manufacture the device at one-fifth the cost, maybe less. The device would produce high voltages, but power electronics would be used to match up with certain products that operate at lower voltages. "It's really tuned to the electronics we attach to it," Weir once told me. "We can go all the way up — from pacemakers to locomotives to direct-energy weapons."[6] The power electronics would also take what in an ultracapacitor is usually a quick release of energy and reduce it to the kind of steady flow of current we get from batteries. If it all works — and there are many industry experts who believe Weir and Nelson will never get it right — this is the kind of "fundamentally new approach" that Khosla thinks is needed to dramatically open up the battery bottleneck.

HIGH-RISK BET

How much of this ceramic battery technology is based on Weir and Nelson's top-secret work for TRW, or materials and thin-film manufacturing processes developed at Titanium X? Only they know. But once the core patents were filed, Weir began hunting around for investors in 2002. Sources say Weir approached some of the big automakers and was promptly rebuffed. They had little interest in electric cars at the time, let alone a vehicle powered by a little-understood energy storage technology with no track record. They'd rather watch and wait, then pounce later if necessary. It was around the same time that a small Toronto-based company called Feel Good Cars was making its U.S. debut at the Electric Transportation Industry Conference in Florida. Feel Good Cars launched a low-speed electric vehicle called ZENN, which stood for "zero emissions, no noise." In many ways, it was a pimped-out golf cart. It ran on lead-acid battery power but was fully enclosed, heated, and came with air conditioning. It had all the necessary lights and safety features as well as power windows, keyless entry, and a slick stereo system — a perfect ride within a gated community. Weir noticed the media exposure Feel Good Cars was getting, so in late 2002 he reached out to Ian Clifford, the co-founder and at the time chief executive of this small but seemingly innovative electric-car venture.

"We talked quite a bit by phone, and eventually we met in Montreal," said Clifford, a former photographer and Internet marketing entrepreneur who has a personal love of electric vehicles. Clifford was intrigued by the potential of Weir's work and spent much of 2003 leading due diligence on EEStor and its energy storage claims. Interestingly, the references Weir provided to Clifford included contacts at NASA and military contractors Lockheed Martin and General Dynamics, indicating that Weir and Nelson were still working on U.S. defense contracts. Indeed, in one interview, Weir conceded to me that EEStor had been working closely with Lockheed Martin since 2001. Clifford knew that EEStor was a high-risk bet, but he also understood the widespread implications if the company delivered. Partnering with

EEStor would be a leap of faith, but Clifford happily took the plunge. Feel Good Cars struck a $2.5 million licensing agreement with EEStor in 2004 that gave the Canadian company exclusive worldwide rights to purchase EEStor's super battery for use in compact and subcompact cars — pretty much any vehicle weighing up to 1,200 kilograms, excluding small suvs and pickup trucks. "Nobody was talking about electric cars then," Clifford recalled one morning over breakfast at a Toronto diner. "There's a time and a place for everything, and we happened to be in the right place at the right time." A later amendment to their agreement increased Feel Good Cars' market to any car size up to 1,400 kilograms. The $2.5 million was a small price to pay for the potential, however fragile the reality, to dominate a multibillion-dollar vehicle market down the road.

Weir's quest for funding didn't stop there. Through the high-tech community in California, he managed to link up with Ed Beardsworth, who made a living at the time scouring national research labs for emerging technologies that his clients, a group of large power companies, might find a threat or opportunity. Beardsworth was essentially a technology scout for utilities. He knew the energy sector intimately and reported back his findings through his self-published newsletter *Utility Federal Technology Opportunities*, or UFTO. "Once I got to know Dick and his story, he authorized me to connect him with investors," explained Beardsworth, who was based in Silicon Valley and very much plugged into the local venture capital scene. Introductions were made with a number of venture capital firms, including Kleiner Perkins, which at the time had a mandate to invest more aggressively in "green technology" startups. Kleiner was captured by EEStor's story and began its normal due diligence, which included a call to Weir's former Titanium X investor Gary Hultquist, who told them "to not underestimate Dick and Carl, because these guys were very good at what they did." Like Feel Good Cars before them, the high-profile venture capital firm decided in 2005 to take the plunge and led a $3-million investment in EEStor that gave it a 32 percent stake in the company. To keep tabs on Weir's progress, Kleiner placed two big guns on EEStor's board: private-equity investor Morton Topfer, a veteran high-tech executive who was a former vice chairman of Dell Computer and widely known as Michael Dell's mentor; and Michael Long, a seasoned executive with close ties to Kleiner.

Up to this point, there was little public knowledge of EEStor's existence, let alone what the company was all about. It wasn't until privately held Feel Good Cars (renamed ZENN Motor Company in 2007) had a public share issue in February 2006 and began trading on a Canadian stock exchange that information about this secretive company from Texas began to trickle into the public domain. A month later, my first major feature about EEStor was published in the *Toronto Star*, after which the mainstream media began tracking the story more closely, and the volume of chatter in the online investor universe rose. Postings on my blog, cleanbreak.ca, about EEStor attracted thousands of obsessed readers and hundreds of comments. But it wasn't until 2007 that interest in EEStor truly exploded, and the story took on a life of its own, one that was, at times, quite dramatic. Online reports of EEStor, in the blogosphere and on traditional media sites, became lightning rods for largely anonymous speculation and debate about the company's technology and its chances of commercial success. Every statement from Weir or Clifford was seized on, parsed, and analyzed for its significance. EEStor's technology patents were dissected and challenged. Some called EEStor a scam and Weir a snake-oil salesman, while others preferred to keep an open mind and cheer on the company as it attempted to achieve what many branded unlikely, if not impossible. There were both boosters, who owned shares of ZENN and were trying to influence the stock, and bashers, a mixed bag of engineers and scientists suffering from technology envy (what some might call "not discovered here syndrome") or mudslingers paid by EEStor's competitors to discredit Weir by trash-talking his work. After all, EEStor wasn't just threatening the fossil-fuel industry; it was threatening all other high-tech ventures out there hoping to cash in on demand for better energy storage.

Between those two extremes was the mushy middle of true believers and skeptics who followed EEStor's story like it was a television soap opera.

LEAP OF FAITH

Michael Blieden became one of the followers. The Californian documentary filmmaker got hooked by EEStor's story in 2008. He quickly grasped the enormous global implications of what Weir and Nelson were trying to achieve, but he was just as intrigued by the volume of online buzz that had developed around this strangely named company that demonstrated no desire for the attention it was getting. "The degree to which people, with little to go on, invested emotionally in this technology fantasy was fascinating to me," said Blieden, who became part of the growing legion of EEStorians. The story was so unusual, so interesting, that he decided in August 2008 to make a documentary about this EEStor phenomenon. Blieden spent five months and his own money traveling the continent researching and conducting interviews with scientists, investors, bloggers, journalists, and other individuals caught in EEStor's web of intrigue — myself included. "My angle was to tell the story of the technology," he said, "what it could do, but also to put on display the emotional conviction of the people who cared, while also showing my own transformation into one of those people. Based on what? I'm an atheist. I reject the god concept because of lack of proof. Yet here I was doing all of this, without any proof, because of a battery technology. It just didn't make any sense. I think a small part of it is the will to believe it."

By this time, the story had gotten juicier. In January 2007, the company put out its first press release, which was filled with all sorts of technical information about the "purification, concentration, and stability" of the chemicals and materials used in its process, all of it certified by an independent third party. The next milestone, it said, was to get certification of the barium titanate powders used in its final product. "EEStor Inc. remains on track to begin shipping production 15 kilowatt-hour Electrical Energy Storage Units (EESU) to ZENN Motor Company in 2007 for use in their electric vehicles," according to the statement.[7] I managed to convince Weir to chat about the press release for an article I was writing for the website of MIT's respected journal *Technology Review*. Weir didn't mince words. "This announcement says we've not only met our goals, but we've beat them," he said. "We've achieved something dramatic, and there's no bullshit associated with it." I asked him if there was too much hype around EEStor and its technology. "I don't hype things," he snapped back. "We're not a hype company, and we're well on our way to doing everything we said." And what was his long-term plan? "To make North America the energy capital of the world."

There it was for the world to see: EEStor would deliver a working EESU to ZENN by the end of 2007. Expectations for all observers were now at an all-time high, and ZENN's Clifford was stoked. "That's thrilling for us," he told me at the time. "We can't wait. We're first in line. We're the first guys in. We took the initial risk and really believed in what they are doing." Buoyed by the update, ZENN negotiated another deal with Weir that three months later would see it invest $2.5 million in EEStor, giving the company a 3.8 percent ownership in the firm. ZENN also had the option of investing another $5 million after EEStor certified the permittivity of its barium titanate powders, which made up the dielectric in an EESU.[8] (Permittivity levels indicate the amount of charge that can be stored in the device without letting current leak across the two ultracapacitor plates.) But one had to wonder whether EEStor and ZENN were getting ahead of themselves.

That was Andrew Burke's view. A researcher at the University of California–Davis and recognized expert on energy storage technologies for the transportation sector, Burke was among the many who continued to express both concerns about EEStor's claims and doubts about the company's stated commercialization schedule. It is one thing to make ceramic powders and prototypes in a controlled laboratory, he said, but quite another to churn out reliable, safe, and long-lasting energy storage devices on a production line at low cost and high quality. There is a long history of companies that have tried and failed repeatedly to bring such innovations out of the lab and into the real world. Why would EEStor be any different? "It's interesting that they're making progress," Burke told me. "I have no doubt they can develop that kind of material, and the mechanism that gives you this kind of energy storage is clear. But it's a big step from making powders to making devices." And even if EEStor could deliver a product to ZENN by the end of 2007, the idea that EESU-powered cars could start hitting showrooms shortly after was naive at best. What about warranties? Road tests? All the regulatory hoops to jump through? The path to commercialization is much longer and more complicated than simply getting delivery of product. If these realities didn't bolster skepticism of EEStor, then its failure to deliver a product to ZENN by the end of 2007, as Weir had promised, certainly did. The company didn't even meet its "permittivity" milestone by the end of that year. There would be no year-end fireworks for the EEStor faithful.

BEING DICK

As expectations outside the company were rising, there was much infighting going on behind the scenes. Kleiner Perkins, which as a highly influential venture capital firm was accustomed to being heard and having its way, grew increasingly frustrated with Dick Weir's stubborn management style. "Weir is very definite about how it's going to be, and he's determined to not be anybody's fool, or tolerate anybody having influence over him," explained Ed Beardsworth, the man who first connected Kleiner with EEStor. Kleiner was unhappy with EEStor's relationship with ZENN and tried to pressure Weir — through board representatives Mort Topfer and Michael Long — to wiggle out of its agreement with the Canadian company. The venture capital firm felt that ZENN was getting too sweet a deal through its exclusivity agreement with EEStor and that Weir had negotiated away too much, too early. Weir, on the other hand, was determined to honor the deal with ZENN and was put off by Kleiner's attempts to micromanage his business, or, as one source put it, "having the suits looking over his shoulder." Kleiner brought experience and connections to EEStor and had the potential of opening many doors, but Weir far too often rejected the firm's advice. "He would say, 'I know what I'm doing, thanks for the money, I'll let you know as things progress.' Kleiner didn't like that approach at all," according to one insider. "Weir doesn't have any problem at all with maintaining a distant relationship with people, and he's not afraid to tell people to take a hike."

Hoping to get their side of the story, I reached out to both Kleiner and Topfer and received no response. But here is what I managed to cobble together through well-placed sources: Kleiner ruffled some feathers when it decided to go behind Weir's back and sell a portion of its preferred shares to Topfer and Long, giving Kleiner a full return on its initial investment. This little side deal put Kleiner in a comfortable position. It still had much upside if EEStor hit it big, but it also faced little, if any, risk if EEStor crashed and burned. The relationship got strained even further after Weir struck a deal with ZENN as a way to guard against a takeover by Kleiner, which had an option to buy more shares that could have led to a controlling stake in EEStor. This transaction with ZENN, as one insider put it, "was the final nail in the coffin" for Weir's relationship with Kleiner. The venture capital firm still owned about 20 percent of EEStor as of year-end 2010, but it's now sitting on the sidelines as a more passive investor. "We stay friends," Weir once said. I'm not sure Kleiner sees it this way.

Throughout this, Mort Topfer became tired of being a ping-pong ball between EEStor and Kleiner, so the former vice chair of Dell Computer decided in mid 2007 to temporarily step down from EEStor's board. Topfer, it seems, struggled with what he felt was a personal conflict of interest. The Topfer and Weir families know each other well. Dick Weir's sons, Tom (now general manager and vice president of operations at EEStor), Richard, and Greg worked at Dell in Austin with Mort Topfer's sons, Alan and Richard. At the same time, Topfer was a representative of Kleiner and had a responsibility to look out for its interests. The safest thing to do was eliminate himself as the middleman and wait until the dust settled. And eventually it did. Weir put out a press release in January 2008 to announce that Topfer was back on EEStor's board. It was widely rumored that Topfer was helping Weir establish a licensing deal with Dell Computer, which would use its exclusive access to EEStor's energy storage technology to gain an edge over rivals, particularly lower-cost competitors from China. There was no substance to the rumors, however, other than Topfer's close connection to Dell and the fact that such a deal would make great sense. (Sometime in spring 2009, Topfer, who was 72 at the time, once again resigned from EEStor's board, leading to speculation that the company was having problems with its production process. But Topfer may have had other reasons unrelated to EEStor's performance, as he also left the board of computer chip giant Advanced Micro Devices just a few months later.)

Weir was on a bit of a roll that January, and that set the tone for much of 2008. The same day that he announced Topfer's return to EEStor's board another press release came out, this time from military contractor Lockheed Martin. The $40-billion behemoth, which employs 140,000 people around the world, disclosed that it had signed an exclusive international rights agreement with EEStor. The deal gave Lockheed Martin the license to integrate and market EESUs for any military and homeland security applications — sensors, weapon systems, vehicles, basically anything that requires power and mobility on the battlefield. I spoke that same day with Lionel Liebman, a program manager who oversees applied research at Lockheed's missiles and fire control division. "I think it's very real," he said of EEStor's technology. Liebman added that he had visited the company's facility in Cedar Park, Texas, and was impressed with the progress. He went on to talk about the challenge of getting power to soldiers in the field, relying less on conventional fuels and using more renewable energy in a military context. EEStor's EESU could be a game-changer, he said, adding that Lockheed would work closely with the company to build new prototypes. One project Lockheed was deeply involved in was the development of rechargeable, light-weight body armor for soldiers that could double as a power source for in-field weapons and other portable gadgets, or even biometric sensors that could monitor and transmit soldiers' vital signs in real time. In other words, the armor would be a wearable battery designed to protect soldiers, who are less nimble in combat when required to haul around 10 or 20 pounds (or more) of lithium-ion batteries. In a "body armor" patent filed later that year, Lockheed specifically mentioned EEStor as a possible source of power storage for the ruggedized military vest.

Weir the patriot, the Marine jetfighter pilot, the man who worked on top-secret projects for the Star Wars program, must have felt in his element. "I'm really in deep with Lockheed," he is heard saying on his leaked interview, and this relationship with Lockheed appeared to be paying off. But Lockheed's involvement raised a new batch of questions. Did EEStor have a working prototype? Did this new Lockheed distraction explain why EEStor didn't deliver product to ZENN by the end of 2007, as initially promised? Was Weir holding back on something? Why would an industrial giant like Lockheed go out of its way to put out a press release? The rumor mill continued to grind and conspiracy theories began to flourish. It seemed the more information that was revealed about EEStor, the greater the mystery surrounding the company and the more polarizing it became. When would the world get to kick the tires on a real, working product?

WAITING GAME

All of this built-up anticipation seemed to make Dick Weir crankier than usual. He didn't like having to defend EEStor or explain its delays, or having his work — at least what was known through his many patents — doubted or misinterpreted. The speculation bothered him, but his unwillingness to open up made speculation the only option. This is the conundrum for anyone working on disruptive technology: be too transparent and people will question what you've got; be too closed and people will question what you've got. It can be a no-win situation, as speculation equals risk, and too much risk makes it more difficult to attract strategic partners and the kind of funding necessary to take the big leap from R&D to commercial production. Miss too many deadlines without explanation, and people — from anonymous bloggers to potentially important customers and suppliers — are bound to suspect that something is wrong, even though such delays could have nothing to do with the technology. EEStor's dysfunctional relationship with Kleiner didn't help matters, as word of such tensions spread quickly through business circles. "Frankly, I couldn't tell you if EEStor is a bad idea, a good idea, or a fraudulent idea," venture capitalist and former Kleiner partner Vinod Khosla told me in a conversion in spring 2010. "My bet is it's probably a good idea that needs more help; that the founders don't know what to get because they *think* they know what they're doing." Khosla, still an affiliated partner with Kleiner, has always said he is "somewhat skeptical" of EEStor but has never denied that the science behind its technology can't be made to work.[9] All of this is a reminder that good technology alone, however disruptive its potential, does not determine success or failure in a market with many moving parts and self-interested players.

The next time I had a chance to speak to Weir was in July 2008, after he announced a milestone was met, originally scheduled for 2007. EEStor disclosed that it was producing the materials in its EESUS — the modified barium titanate powders and aluminum oxide particle coatings — at high enough purities to tolerate the extreme voltages the device was designed to handle. As with the previous milestone, the claim was certified by Dr. Edward Golla, laboratory director for chemical testing facility Texas Research International, who appeared to be moonlighting as an independent consultant. I asked Weir when a working product would be shipped to ZENN. "I'm not going to make claims on when we're going to get product out there; that's between me and the customer," he shot back, but he also spoke like a man on the cusp of something big. "It's the home run," he said. "All the key production items have now been certified to go into production, and all the equipment to do it is sitting in our factory." The next step sounded like it would be major expansion; no longer was EEStor doing R&D or pre-production. From this point on, it was about high-volume production, which he wanted to "wrap up" by the end of the year. "Like shit going through a goose," he told me. "Now it's just turning a production crank and putting the machinery in." He indicated he was working 16-hour days, seven days a week and staying up late at night hammering out patents, and labeled the whole effort as his "Manhattan II" project. Soon he would certify permittivity, then an EESU component and then a full EESU. "After that, we'll all go to Tijuana and get drunk for a month." Was he worried about competition? "If we get challenged, we'll move to scale up. We've got a lot of knowledge built up."

That was a lot for the online community to chew on, and it was around this time that Internet chatter reached a fever pitch. In 2007, much of the online debate took place on blogs, such as my Clean Break blog or the GM Volt blog, established by electric-car enthusiast Dr. Lyle Dennis as a way to track progress on General Motor's much-anticipated Chevrolet Volt. EEStor followers also gathered in the comment areas on news sites, such as MIT *Technology Review*, which had published articles about EEStor, or in Yahoo and Stockhouse investor billboards. Noticing all of this online traffic, an anonymous blogger based in the Washington, D.C., area decided get in on the action. Known only as "B," he created the Barium Titanate blog and later built it into a dedicated website, TheEEStory.com.[10] "There are a few thousand who read the site each day, and on [days with] news, it can get close to 20,000 unique visitors," said B. The site quickly became a sort of Wikipedia dedicated to all things EEStor. Ed Beardsworth, who has tracked EEStor's progress since he first introduced Kleiner to Weir, said the site's several writers and regular visitors dig up everything, including patents, timelines, background information on the cast of characters involved with EEStor, and any story that mentions the company. Some have gone as far as submitting freedom of information requests to the U.S. government, including to the Pentagon, seeking access to letters and memos that mention EEStor. They even do original research, including interviews with "people of interest." What they get back, they immediately post on the site. "There must be something the size of *Encyclopedia Britannica* assembled in that EEStor archive," said Beardsworth.

Michael Blieden, the documentary filmmaker, was also on top of the story. "I went deep," he said, having convinced himself that EEStor, even if not ultimately successful, couldn't be a scam. Dick Weir and Carl Nelson were experts in what they were doing. The relationships with Kleiner and Lockheed were far too compelling to ignore. Blieden had heard rumors that the big announcement — the big product reveal — would now be taking place in March 2009, so he packed his gear, kissed his wife goodbye, and drove from California to Austin, Texas, where he sublet an apartment. "I thought I needed to be there to capture the moment." But life, as it so often does, got in the way. Blieden received a call at the end of January from the makers of *Late Night with Jimmy Fallon*, a new comedy-talk show on NBC. It was a big break for Blieden, so he temporarily shelved his documentary and moved to New York. "This project is still very close to my heart," he said. "It's still my number two priority." It was a good move in retrospect, as he didn't miss much. March 2009 came and went with nothing but silence from EEStor.

There was some progress made through 2009, even if climactic expectations weren't met. EEStor announced in April that the permittivity tests of its chemical powders were certified and exceeded targets, meaning its device was capable of meeting the company's claim of storing enormous amounts of electrostatic energy. After the achievement of this milestone, ZENN exercised its option of investing another $5 million in EEStor and upping its stake in the Texas firm to 10.7 percent. Then, in July, the 30-minute conference call between potential investors and Dick Weir mysteriously made it onto YouTube. In it, Weir talked confidently about his EESU "making wind farms operate like coal plants" and offering "three to five times more energy storage" for laptop computers and handheld devices, which would "never degrade" and "charge in seconds." He said he was "knee deep" with people in the portable tool industry who were waiting anxiously for EESUS to emerge. Regarding electric vehicles, he said ZENN would get its EESUS at a cost of $100 per kilowatt-hour, excluding associated electronics, compared to $350 to $1,200 for competing lithium-ion technologies. "Nobody is going to compete with us, certainly no lithium-ion," he insisted. Weir, having failed to deliver in 2007 and 2008, also offered a new forecast. "I'm already out there putting EESUS together, and I'm still in June. I'm ahead of schedule," he said, seemingly oblivious to how much the schedule had changed over the years. He said ZENN would get pre-production prototypes by the end of 2009, at which time "all hell is going to break loose" heading into 2010. Was this all an embarrassing disclosure of a confidential chat, or was it a clever stunt? Like a never-ending soap opera, EEStor, even if unwittingly, continued to deliver new plot twists without actually delivering product. By the end of 2010, there was still no EESU.

FOOL ME ONCE . . .

Energy blogger Charles Barton once wrote that EEStor will turn out to be one of two stories: "The first will be a story of great deception, with self-deception perhaps playing a key role. The second would be of a great technological triumph by brilliant American inventors, Texans who made oil obsolete." Needless to say, in 2010 the group who thought EEStor a scam or a misfired missile was growing, and the believers were beginning to lose their faith. The leaked conference call seemed to plunge EEStor into darkness, and ZENN stopped speculating about delivery dates, likely because securities regulators in both Canada and the United States slapped their hands. What ZENN did announce in September 2009 was a gradual shifting of its business strategy, presumably in an effort to stay focused and conserve cash. Where it once had grand plans to make its own EEStor-powered car, the cityZENN, it was now going to concentrate efforts on developing an electric-vehicle drive train, dubbed ZENNergy, which would be based on EEStor's technology and supplied to customers, namely the major automakers. In other words, little ZENN, instead of getting into the capital-intensive business of making cars, would pursue an "Intel-inside" business model. ZENN also said it was discontinuing production of its existing low-speed electric vehicle and getting out of that market entirely. Its future, for better or worse, was now in EEStor's hands.

Ian Clifford, shortly after ZENN announced its transition plans, did a noble job of hiding his own impatience while sipping coffee one fall morning in 2009. He was anxiously waiting like everyone else, as EEStor is not obliged by any contract to meet a specific delivery date. But the difference for Clifford, unlike everyone else, was that the future of his company was directly tied to the outcome. "Obviously, our entire business model is dependent on EEStor commercializing that technology," he told me, choosing his words more cautiously than in the past. There were no more public mentions of targets and deadlines and schedules. "Delays are frustrating to the extent that the whole world is waiting. Internally, I know what Dick is working on and why he's taking the time he's taking, but like anything disruptive, it can't be happening soon enough," he said. "We remain optimistic. The transformative moment is with the commercial proof, and when that day comes the whole tenor of the discussion changes to excitement about the reality, not just the promise."[11]

The outside world, however, isn't sitting idly by waiting for EEStor to come through. The longer Dick Weir takes, the more opportunity there is for competitors to close the gap. How quickly is the gap closing? I posed this question to Dave Pascoe, vice president of the electric-car systems division at auto parts giant Magna International, as we sat in Magna's cafeteria. "It's a very active area right now," said Pascoe, a tall and lean man who looks a bit like a 40-something Jimmy Stewart. "A lot of companies recognize this is a big opportunity if you can come up with a cost-effective, high-energy, high-power storage device."

Pascoe is an open-minded guy who likes to collect quotations from interesting people. On his BlackBerry device, he pulled up a memorable quote he'd solicited during a meeting with U.S. inventor extraordinaire Dean Kamen, founder of DEKA Research and Development Corporation and creator of the Segway Personal Transporter. It read: "School teaches you how to fail; it doesn't teach you how to succeed. To succeed you have to be willing to fail. You have to be willing to be ridiculed and the ridiculers will be right and it will hurt." Pascoe is exposed to all sorts of new technologies kept from the public eye, and he knows that failure is a necessary part of invention and innovation. It's for this reason, perhaps, that he talked with a sense of hope as he recalled (but wouldn't name) the many bleeding-edge storage technologies he has seen. "We won't necessarily see a 10-times improvement, but there are certainly game-changers out there," he said. I asked him about EEStor. "Quite honestly, I hope they're successful, but I still only give them a 5 percent chance." Until the game-changers come, Pascoe figured plug-in electric vehicles will be limited to capturing 1 percent, maybe 2 percent of all new vehicle sales by 2020. Some experts are more optimistic. Sarwant Singh, vice president of the automotive technology practice at research firm Frost & Sullivan, predicted in 2009 that within 10 years hybrid and purely electric vehicles will account for 7 to 12 percent of all new cars globally, with Chinese manufacturers representing nearly half the market.[12]

The Chinese are forcing a worldwide acceleration of energy innovation, a growing concern for U.S. officials who fear their country is falling behind. U.S. Energy Secretary Steven Chu has said that America faces a "Sputnik moment" when it comes to clean energy — a reference to the Soviet Union's 1957 satellite launch that triggered a Cold War space race and led to an era of unprecedented technology innovation in America. Speaking in late 2010 at the National Press Club in Washington, D.C., Chu touted some made-in-America companies that could put the United States back on track. He specifically mentioned Arizona-based Fluidic Energy, which is working with Arizona State

University on a new type of "metal-air" battery capable of "carrying a four-passenger electric car 500 miles without recharging, at a cost competitive with internal combustion engines." Fluidic and Arizona U are aiming for energy densities of at least 900 watt-hours per kilogram and up to a stunning 1,600 watt-hours per kilogram — three to five times what EEStor claimed in its original patents.[13] (In September 2010, EEStor amended one of its patents to indicate it can achieve a three-fold increase in energy storage over its original claim, putting it in line with the energy densities being pursued by Fluidic Energy.[14])

Interestingly enough, while critics continue to dog and doubt EEStor, there is growing interest in academic circles of the potential of ultracapacitors as a replacement for, not just a complement to, electrochemical batteries. In April 2010, for example, the U.S. Department of Energy granted tens of millions of dollars in funding to high-risk, high-reward energy storage research initiatives, including one ultracapacitor project at Penn State University that's aiming to design a "cost-effective alternative to battery solutions" with potential for "higher cycling" and "high power density." Penn State, known globally for its expertise in materials science, is partnered in the project with Recapping Inc., a stealthy company that is headed by Alex Kinnier of Khosla Ventures. Vinod Khosla has always had a cautious fascination with EEStor, and it now appears he's backing an EEStor-like company of his own. He denies it, of course. In an email, Khosla described Recapping as a "science investigation" that shares little in common with EEStor. Still, one could argue that EEStor, win or lose, has inspired such investigations and brought consideration to a technological pathway historically written off as a dead-end pursuit. The difference is that EEStor, unlike Recapping and Fluidic Energy, is well past investigation and R&D. In Weir's mind, commercialization is the next stage and EEStor's impact on the world is imminent, even if the view from the outside isn't as rosy.

THE ROAD AHEAD

Nearing the end of 2010, the popular green-technology news site GreentechMedia.com published an article titled, "Is the EEStor Saga Finished?" It was a fair question, considering, as the article pointed out, there was still "no news, no website, no chatter from the ultracapacitor urban legend except for some government skepticism."[15] The government skepticism the article referred to related to U.S. Air Force Research Laboratory documents obtained through freedom of information requests.[16] In them, various individuals at the lab, whose names were blacked out, expressed their doubts about EEStor. One individual said EEStor was "full of it" and another, while recognizing the potential of the technology to revolutionize many energy and technology applications, advised lab staff to distance themselves from the company. Their opinions weren't based on what they knew; they were based on what they didn't know. Again, Dick Weir's tendency to hide his cards and put on a poker face hadn't helped matters. But when you're sitting on a potentially world-changing technology, how much is safe to reveal? Clearly, Weir has an interest in protecting his intellectual property and remaining ahead of potential rivals, including the Chinese, for as long as possible. "Where's the happy medium?" asked Ian Clifford, who is sympathetic to Weir's dilemma. "It's really hard to find that balance. And really, show me a disruptive technology that doesn't have a polarized audience. It comes with the territory." It may be that no amount of proof, save a mass-produced working product, will satisfy the skeptics.

Kleiner Perkins, it should be noted, it still watching EEStor closely and has not written off the company's chances just yet. That was the impression Bill Joy left while speaking in April 2011 at a technology event hosted by the Massachusetts Institute of Technology. Joy, a partner in Kleiner Perkins' green technology practice and co-founder of computing giant Sun Microsystems, is known for having stickhandled the venture capital firm's early investment in EEStor. Joy has kept quiet about EEStor for several years, but when asked at the event if he still had confidence in the company he replied, "Oh, yeah," at the same time emphasizing the difficulty of its mission. "What they are proposing to do is wild. And there are lots of reasons some of these things could fail to be commercialized. . . . But worthwhile things usually are hard, and they always take longer." Joy said EEStor's capability to deliver directly relates to the knowledge that Carl Nelson has brought to the venture. "I mean, if you ask Carl, tell me something about hafnium, Carl will talk to you for 15 minutes about hafnium. He knows something about every element in the periodic table. He can tell you off the top of his head what its crystal forms are and other interesting properties. . . . You just have dinner with him and you realize he really understands what he is doing with materials." As for Dick Weir, Joy neglected to mention his name.

EEStor may have already had its coming-out party by the time you read this. Or maybe it has filed for bankruptcy. Or perhaps nothing much has changed — we could all still be waiting, debating, scouring, analyzing, encouraging, and trash-talking what may shape up to be the story of the century or the dud of the decade. Whatever the outcome, the world envisioned by Dick Weir will eventually come, even if Weir isn't the one to deliver it. We will have electric cars and scooters and bicycles that can travel hundreds of miles on a single charge and which middle-class families can easily afford. We will have an electricity system supplied increasingly by solar and wind energy that can be stored inexpensively and tapped when we need it. It's a question of when, not if.

Energy transitions have historically been painfully slow, and even disruptive technologies can still get bogged down along the path to commercialization. There are simply too many powerful interests, established industries, standing in the way. But if it wasn't for people like Weir, those mad-as-Tesla characters who put it all on the line out of a belief that they can change the world for the better, we would never get to where we need to be. Weir isn't "full of it" — nobody who can bring big names like Lockheed Martin, Kleiner Perkins, and Mort Topfer to the table is. He's just a guy giving it his own personal moon shot. Good on him, you might say — at least he got this far.

Thane Heins, as you'll read in the next chapter, hasn't made the same progress. Far from it. But that's what happens when the breakthrough you claim challenges the very laws of physics.

Notes:

1 Justin Hibbard, "Kleiner Perkins' Latest Energy Investment," *Businessweek.* September 3, 2005.

2 This rough analysis does not include the weight of any power electronics EEStor would need to manage power output of its system.

3 On July 21, 2009, a mysterious 38-minute audio interview with Dick Weir was uploaded to YouTube. It was later discovered that the interview was arranged and conducted by Paradigm Capital, a Toronto-based investment firm. Paradigm interviewed Weir as part of due diligence for investors in ZENN Motor, a publicly traded company with investment in EEStor that was in the process of raising more money from a public share issue. Paradigm claimed it did not know how the interview leaked. (See: Tyler Hamilton, "Paradigm Capital Red-Faced After Conference Call Posted on Net," *Toronto Star.* August 8, 2009.)

4 Vinod Khosla, "The Limits of Today's Electric Car Technology," *Grist Magazine* online, August 9, 2009.

5 "Atmospheric Electricity," Meridian International Research. http://www.merdian-int-res.com/energy/atmospheric.htm.

6 Tyler Hamilton, "Battery Breakthrough?" *Technology Review.* January 22, 2007. http://www.technologyreview.com/business/18086/.

7 "EEStor Announces Two Key Production Milestones: Automated Production Line Proven and Third Party Verification of All Key Production Chemicals Completed," press release, EEStor. January 17, 2007.

8 "ZENN Motor Company makes equity investment in strategy partner, EEStor Inc.," press release, ZENN Motor. April 30, 2007.

9 Vinod Khosla.

10 The blogger "B" says he created the blog in 2007, though there is one entry dated 2004. He wouldn't explain why, so I assume he backdated the entry. In any event, postings and traffic on the site didn't really get going until fall 2007.

11 On February 14, 2011, ZENN announced that Ian Clifford was stepping down as CEO and taking on the role of vice chairman to help the company preserve cash. His new role would be dedicated to maintaining ZENN's relationship with EEStor, according to a company press release.

12 Sarwant Singh, "Electric Vehicles Unplugged: A 360 Degree Global Perspective," presentation to Electric Mobility Canada. September 29, 2009. (Singh works at Frost & Sullivan.)

13 Tyler Hamilton, "Betting on a Metal-Air Battery Breakthrough," *Technology Review*. November 5, 2009. http://www.technologyreview.com/energy/23877/.

14 Susan Taylor, "Zenn Shares Rises as Partner Amends Battery Patent," Reuters. September 30, 2010.

15 Eric Wesoff, "Is the EEStor Saga Finished?" GreentechMedia.com. December 1, 2010.

16 It seems the freedom-of-information requests were submitted, received, and posted online by "B," the mysterious blogger who runs TheEEStory.com.

CHAPTER 7

Searching for Miracles

Changing the World with an Open Mind

"I have learned to use the word 'impossible' with the greatest caution."
— Wernher von Braun, rocket scientist

Ed Beardsworth, the technology scout who introduced Dick Weir to Kleiner Perkins, began what he considered to be the world's best job in May 2008. He was approached by a "high net worth" individual who was part of an investor group with major energy and other assets throughout Asia. Like most of us, this individual wasn't happy with the state of the world. Climate change. Pollution. Energy security. Energy scarcity on an overpopulated planet. You name it. He wanted Beardsworth to oversee a project, dubbed The Hub Lab, which would strive to identify and fund the next big discovery in energy physics, the kind of taboo science historically dismissed and derided by the mainstream scientific and investment communities.

The ideas and technologies discussed so far in this book

would seem safe bets by comparison, despite the enormous skepticism and many challenges they face. Whether a technology can be developed cheaply enough, is safe, can scale up, is based on materials that are widely available, can attract capital, can be made to work as envisioned, or is compelling enough to take on established industries or completely overhaul supply chains — those are some ways of measuring what's commercially possible, of determining chances of success or failure. What Beardsworth was seeking out, however, were potentially disruptive technologies that, at their core, were believed to be scientifically impossible. His job was to tap into a vast underground network of scientists and inventors working at the fringe, get to know them, learn about their work, and then supply funding to the handful who showed the most promise. The project lasted two years and, in all, Beardsworth handed out seven grants.

"It was more fun than anybody should be able to have," recalled Beardsworth, who looked at everything from cold and aneutronic fusion to so-called free energy devices to systems that harvest zero-point energy. Much of what he saw had no backing theory and some projects claimed to defy known laws of physics. Figuring out which ones to fund was no easy task. "We wanted fringe science, but we didn't want the wild-eyed, naive nut jobs out there, because this just takes you off in incredible directions. You don't know if a person is sane or not, or if they have any clue what they're talking about," said Beardsworth. "Because people are stretching the limits and coming up with new ways of seeing the world, you don't necessarily have a ready way to evaluate it." Much of the time, he said, it came down to what the person was like, not just the technology. "We cared as much about the personality and work style of the person we gave grants to as we did their ideas. Is it somebody who's going to keep a careful lab notebook, properly record measurements, show data, and be methodical? The problem is that 90 percent of the people are probably not worth the time of day. The other 10 percent that are competent and grounded get lost in the noise and tarred with the same brush."

Do we know all there is to know about everything? Are established laws of physics the last word on science? Is there room for these laws to be broken, amended, or broadened? Is there energy out there that we haven't discovered or haven't figured out how to tap? History has shown us that there is always more to know, and figures from our past, including Nikola Tesla, are held up as proof that there's not just room for unconventional thinking, but that challenging what we know and taking an unorthodox approach has rewarded us before and can reward us in years ahead.

As recently as the early 2000s, the U.S. Department of Energy told Ukraine defector Dr. Valeriy Maisotsenko that the super-efficient air-conditioning system he had developed defied the laws of thermodynamics.[1] As a result, they rejected his request for a development grant. Five months later, Maisotsenko, who lived in Denver, sent the government naysayers a prototype they could test. What they thought was impossible turned out quite real: an air-conditioning machine that used 80 percent less energy than a conventional system. They conceded that Maisotsenko had discovered a new thermodynamic cycle, which has since been named the Maisotsenko Cycle. A company in Denver called Coolerado now sells the groundbreaking air conditioners, which cost about 20 percent more but have an impressive two-year payback.[2] Clearly, there are still many surprises to be discovered.

HYDRINO HOPES

Some scientific observers wonder whether BlackLight Power of Cranbury, New Jersey, has already uncovered another surprise, even if the rest of the world isn't yet prepared to see it. Founded by Harvard-trained medical doctor Randell Mills in 1991, BlackLight claims it has discovered a non-polluting primary source of energy called a hydrino, which it considers a form of hydrogen that is "likely" the dark matter of our universe. Mills believes a hydrino is a low-energy form of hydrogen, meaning that when a hydrogen atom is converted (or shrunk) into a hydrino it releases energy in the process — up to 200 times more energy than burning hydrogen gas directly as a fuel. It's impossible according to established quantum theory, but Mills is looking to rewrite some of the rules and is ruffling more than a few feathers in the process. "His theory really pisses people off, because they don't like their quantum mechanics rewritten by a mere doctor," said Beardsworth.

If Mills is right — there's that big *if* again — the implications are enormous. Already, BlackLight says it has come up with proprietary chemical catalysts that trigger the hydrogen-to-hydrino reaction, which releases tremendous amounts of heat that can be used to generate clean and inexpensive electricity. The initial hydrogen comes from electrolysis of water, and the electricity needed for water electrolysis would represent less than 1 percent of total energy produced from the process, which BlackLight calls catalyst-induced hydrino transition (CIHT). "The process does not give rise to pollution, greenhouse gases, or radiation," according to the company. It envisions making small electrical and heat generators for homes and communities that eliminate the need for the construction of large, multibillion-dollar power plants. Mills figures that the equivalent of 1,000 megawatts, which is about the generating capacity of a current-day nuclear reactor, could be fueled for a year with 32 million liters of water.[3] That might sound like a lot, but consider that a 1,000-megawatt nuclear reactor requires 1.8 trillion liters of water a year just for cooling purposes, of which 18 billion liters is lost through evaporation. BlackLight also sees its process powering spaceflight, airplanes, boats, trains, and as a source of energy for electric cars. "A CIHT electric car is expected to have a range of 1,500 miles on a liter of water," according to the company.

Mills isn't just another garage inventor living his 15 minutes of fame. Mainstream physicists may consider him a kook, but the Harvard doctor, who over the years has pounded out several books and peer-reviewed papers on the subject, refuses to go away and continues to build support for his claims and theories. When he first introduced the concept of hydrinos in the early

1990s, he argued that what Pons and Fleischmann thought was cold fusion was actually hydrogen-to-hydrino conversion — they just didn't know it. Mills also has an alternate theory of the universe based on his discovery of hydrinos that directly challenges the conventional Big Bang theory, which for most physicists is considered settled science.[4] Amazingly, he has persisted. His company, based out of a former satellite manufacturing plant purchased from Lockheed Martin in 1999, has reportedly raised more than $60 million and employs a team of scientists working vigorously on commercial applications of the core science.[5]

There has been some, though evidently not enough, independent validation of what Mills is doing. Since 2008, a team of scientists and engineers at New Jersey's Rowan University, led by chemistry professor Kandalam Ramanujachary, have validated that BlackLight's process works on a small scale in a controlled setting. BlackLight put out a press release in fall 2010 announcing that Ramanujachary's team had validated another company claim: that electricity, not just heat, had been directly generated from a device built around BlackLight's hydrogen-to-hydrino reactions. "With further optimization, there is no doubt that this technology will present an economically viable and environmentally benign alternate to meet global energy needs," Ramanujachary is quoted as saying in the release. "If advanced to commercialization, it would be one of the most profound developments ever."[6] Such a statement could end up a career-limiting move for Ramanujachary, and surely the Rowan University professor knows it. This makes his comment that much more compelling.

Beardsworth, meanwhile, called BlackLight a "mystery wrapped in an enigma" and Mills a "paranoid and secretive" inventor who carefully manages how his story unfolds. "Whether or not he's making excess energy remains to be seen," said Beardsworth. "He hasn't been able to do it consistently, powerfully enough to make a simple tradeshow demonstration of it, but maybe he will. Problem is, he won't disclose his 'secret sauce' and won't let anybody get close enough to assure there are no hidden wires." Mills reportedly has a public demonstration planned for sometime in 2011.[7] Scientists can argue about the underlying theory for decades if they wish. For the rest of us, we just want to see something that works.

A POTENTIAL DIFFERENCE

Unfortunately, there are only so many needles in the haystack, and the haystack is a mountain that keeps getting larger. Just as historical figures such as Tesla can inspire true discovery, they can also fuel a disproportionate amount of delusion. Type the words "free energy" or "perpetual motion" or "over unity" into Google and you'll see what I mean. There's no shortage of folks making extraordinary claims about a new energy invention that can supply the world with an endless supply of clean power or fuel. Entire conferences are devoted to them, such as the annual TeslaTech conference in New Mexico that explores everything from cosmic energy to magnetic healing, or any other technology that allegedly is being suppressed by the government or originated from alien beings. Conspiracy theories abound. "These people think Tesla was on the verge of tapping something magical before he died, and people who want magic in their lives have adopted Tesla," Beardsworth told me. "Whether Tesla lost his marbles [as he got older] or was still a genius with more to give, we don't know. But there's practically a religious fervor surrounding him."

I should point out that journalists like me are magnets for this crowd of well-meaning fringe inventors, and I don't blame them for running to the media. Venture capitalists and bankers won't hear them out. Academics and politicians treat them like lepers. Newspapers and local TV news outlets, on the other hand, are often suckers for the sensational, and media exposure has a way of opening doors — even if just a crack. We've all read the stories, like the inventor in Pennsylvania who claimed to turn seawater into fuel by lighting it on fire with radio waves.[8] In August 2006, an Irish company called Steorn went so far as to place a full-page advertisement in *The Economist* challenging the world's top scientists to test the claim that it had developed "a technology that produces free, clean, and constant energy." The ad quoted a famous line from Irish playwright George Bernard Shaw, "All great truths begin as blasphemies." It's a very powerful and valid statement. But twist Shaw's comment around and it's equally fitting to say most blasphemies lead to nothing great. So far that has been the case with Steorn. Scientists did respond to the magazine advertisement and an independent 22-member jury concluded in 2009 that the company's claim was rubbish.[9]

Here's the problem: we can't set up independent scientific juries every time a company or individual comes forward and makes an extraordinary claim about a technology that can benefit the world. This isn't to say there aren't people out there working on serious science, however unorthodox their direction or unbelievable their assertions. It's also not saying some of these ideas won't or don't work. But we all can't be dedicated to the cause like Ed Beardsworth was in 2008, a seasoned energy technologist with a Ph.D. in physics who was given a bundle of cash and told to go on a wild goose chase. And even he found himself swimming in murky waters.

What makes one extraordinary claim more worthy of exploring than the thousands of other ones out there? How do we sift through the noise and isolate the breakthroughs? How do we keep those rare game-changers from falling through the cracks? I struggle with these questions as a journalist. As an individual with no formal training in science or engineering, I find myself vulnerable to what documentary maker Michael Blieden described as the "will to believe." At the same time, like any banker, academic, politician, or bureaucrat who may want to keep an open mind, I am held back by practical matters. To truly delve deep enough into a claim requires a combination of personal time and financial capital that most of us can't afford, and a level of scientific knowledge most do not possess. For this reason, it's difficult for even the most open-minded observers to go down that rabbit hole.

Not to say I haven't.

In 2007, I received an email from a man by the name of Thane Heins. He identified himself as the founder of a company called Potential Difference that was based out of a lab room at the University of Ottawa. He explained how he had spent the previous two years working on a new type of electric motor-generator he had invented in his basement. He called it the Perepiteia Generator, named after the Greek word "peripeteia," which means a sudden change of events or reversal of circumstances. The device is designed to manipulate a phenomenon called back electromotive force, or back-EMF, which is a kind of magnetic friction that affects the efficiency of electric motors. Put simply, back-EMF is part of what assures that the amount of energy that comes out of the device is less than what goes in. In physics, it is explained under the law of conservation, Lenz's law or the law of diminishing returns, and it can also be explained under Maxwell's theory. Heins claimed he had eliminated that friction, which would make his device 100 percent efficient. But more than that, he said he had figured out a way to redirect the magnetic field so that it could boost, rather than drag, the performance of the motor. If he could make it work in a practical application, it could enable what he called "regenerative acceleration" in electric cars, meaning the vehicle could potentially charge up its batteries, rather than drain them, while accelerating. The faster you drove, the more juice you'd send to your battery pack. It would eliminate the need to plug in your electric vehicle. To charge it, all you would have to do was drive.

I didn't need to be an electrical engineer to know that this was a clear violation of thermodynamic principles. Heins didn't like to use the words "perpetual motion," even if he did allude to it, but clearly he was saying he could harness a source of energy previously unknown or untapped. Normally I would politely brush off this type of claim, but there was a hook to this story. Heins told me he would soon be traveling to Boston to demonstrate his technology to Markus Zahn, a respected electrical engineering professor at the Massachusetts Institute of Technology and a program director at the university's Laboratory for Electromagnetic and

Electronic Systems. If there was anyone qualified to assess Heins' claims it was Zahn, so I figured there was no harm in waiting to hear what the good professor had to say.

THE HEINS EFFECT

I warmed up to Heins, and it became apparent soon enough that this wasn't just a story about an unbelievable claim; it was also about an individual's struggle to be taken seriously. Heins was 46 when we first connected, and he struck me right away as a friendly but straight-talking fellow. He was the classic underdog — mildly dyslexic, weak in mathematics, and an underperformer in school. There was no graduate degree in physics or engineering hanging on his wall. He did once enrol in a community college electronics program in Quebec but never completed it. What he did have was a chef's diploma, and that landed him some cooking jobs before he eventually started up his own restaurant in a small town near Ottawa. He also has strong political opinions and a keen sense of justice, which in some ways explained his motivation as an entrepreneur and inventor. Heins saw fossil fuels as the root of instability in our society and the cause of needless suffering and death. He blamed the terrorist attacks of September 11, 2001, on our economic addiction to oil, and he was driven to make a difference, even if it created havoc in his own life. Indeed, he was so obsessed with developing his generator as a solution for breaking our oil addiction that his marriage fell apart. "I've tried to quit many times," he once confided. "But I had this idea and I believe it works." For this reason, he has persisted when others might have given up, despite the ridicule and despite the number of slammed doors.

The one door that hadn't been slammed was at the University of Ottawa, where Heins was permitted to conduct his experiments. Supervising his work was Riadh Habash, a professor with the university's School of Information Technology and Engineering. A quiet and cautious man, Habash didn't really want to talk when I contacted him, but he acknowledged that Heins was able to cause an electric induction motor to accelerate when it should be slowing down. "But when it comes to an explanation, there is no backing theory for it," he said. "That's why we're consulting MIT." When Markus Zahn did eventually see Heins' demonstration, he emerged somewhat baffled. He told me during a brief interview that he had seen an "unusual phenomena" and that it was a surprise for him. Still, he dismissed any notion that it broke some law of physics. What it did do was leave open the possibility that Heins had come up with a way to make electric motors more efficient. "I saw it. It's real," Zahn said. "Now I'm just trying to figure it out." Developing a more efficient motor wouldn't be an insignificant outcome, he added. "There are an infinite number of induction machines in people's homes and everywhere around the world. If you could make them more efficient, cumulatively, it could make a big difference."

I ended up writing about Heins and his meeting with Zahn in a story that appeared in the *Toronto Star* on February 4, 2008. The response from the public was unexpected. Hundreds of emotional email messages flooded my inbox in the weeks that followed, and the story spread like wildfire through the blogger community, including popular sites such as Gizmodo and Wired. Many messages and blog posts encouraged Heins to march on and demonstrated that powerful will to believe. They praised him for bumping up against the wall of scientific conformity and pointed out how others in the past — Nikola Tesla, Leonardo da Vinci, Galileo, and the Wright brothers, to name just a handful — faced similar dismissals that proved premature. "Anything that makes the scientific community defend their long-held beliefs is healthy," wrote one commenter. "Too often they scoff and dismiss like church scholars being told the Earth was not the center of the universe." But just as many readers ridiculed Heins and painted him as either an obsessive lunatic or a scam artist looking for naive investors. Some also took aim at me and the *Toronto Star* for giving Heins a platform to "mislead" and "confuse" the public. "Shame on you," wrote one commenter. "I don't think it's necessary to lend credence to perpetual motion machine inventors who are probably out to scam money from investors." Needless to say, it was a controversial yarn, and it became the second-most read story on the *Toronto Star*'s website for all of 2008.

Thane Heins demonstrates his Perepiteia Generator at the University of Ottawa in January 2010. The university offered him lab space to develop his technology.

After that, I made a point of regularly checking in with Heins. I was curious to see where all the public exposure would lead him. This, for me, had become a human interest story just as much as a technology story. Proving Heins right or wrong was not my intention, nor was it within my capacity to make such a judgment. I wanted to see the moves Heins would make next as a result of my article, which, rightly or wrongly, put him in the media spotlight for a few more months and appeared to open up some more doors. How, I wondered, would he position the technology as he attempted to take it to the next level? Would he promote it as a breakthrough in energy efficiency, or risk more pushback by claiming he was violating the law of energy conservation by achieving over unity — that is, getting more power out

217

than going in? How would he react to ongoing skepticism? How well would he work with those who offered to help?

Heins had an open-door policy through 2008 and 2009 that attracted a number of curious visitors to his lab for live demonstrations. For those who couldn't attend in person, he created video demonstrations and tutorials and regularly uploaded them to YouTube. It didn't take long before undergraduate students at the university began calling Heins' work area the "Perpetual Motion Lab," a label he didn't seem to mind at the time. Based on emails he shared with me, it appears Heins had serious queries from a number of respected government organizations, including the magnetics lab at the NASA-Goddard Space Flight Center, the U.S. Air Force Research Laboratory and Defence Research and Development Canada. (One scientist there said his interest was purely personal and not related to his role at the department.) But nothing ever seemed to come of it. He was also approached by a number of businesses and organizations looking to collaborate. For example, a generator supplier in California wanted to adapt Heins' Perepiteia technology for use in a new line of products. The partnerships always seemed to start off well and gradually fizzle out. Each time I contacted Heins for an update, he was onto the next project without completing the previous one. He excelled at reaching out to people and was proficient at leveraging his interactions with one influential group or person as a way to establish a dialogue with another.

MAGNA-NIMOUS
One person he reached out to was Dave Pascoe, the mechanical engineer heading up Magna International's electric-vehicle systems division and an individual in a position to help. Pascoe was curious enough by what Heins was claiming that he made at least two trips to Ottawa in fall 2009 to see a demonstration. He had his doubts like anyone else, but was willing to listen and learn. "There are many things in the past that have been disproven. I like to keep an open mind, and that's one of the reasons I went

out there," Pascoe told me several months later. "Most people wouldn't have done it, but for me, there's the potential side benefit that, while he may not be doing over unity on energy, he may have something that improves efficiency. Usually when you see something that's not behaving the way it's expected to, there can be some nice unexpected surprises." And that's exactly what Pascoe observed during his time at Heins' lab — odd motor behavior, signs of improved motor performance, but no convincing evidence that any laws of physics had been defied. His gut told him that Heins was helping an inefficiently run motor operate more efficiently, a theory I had heard before but which Heins summarily dismisses. In fact, one individual who advanced such an explanation was dismissed by Heins as an "ignorant whiner."[10]

Still, Pascoe remained intrigued and wanted to see more data to back up Heins' extraordinary claims. He liked Heins and thought he was both smart and quite aware of the nature of his claims, and Pascoe sympathized with what the inventor was going through. "In his mind, he's proven it to himself. If you put yourself in his brain, he can't understand why nobody believes him." Pascoe suggested that Heins do another, more conclusive test that would more accurately verify or cast doubt on earlier measurements. "In my mind, I was doing him a favor. Disprove it and get on with your life, or prove it and move forward." But Heins responded to the test request defensively. "Unfortunately, Dave has rejected the 'obviousness' of the demos," he wrote in an email at the time. "Magna wants these tests to compensate for their own lack of trust in the tests we have already done." He seemed willing enough to do the test but said he didn't have the $25,000 required to purchase a measurement device called a torque sensor, which Pascoe considered crucial for a proper evaluation. Nor did he have access to one. So the new tests, to my knowledge, were never conducted.

I asked Pascoe if he was concerned that he may be letting a potentially beneficial, if not disruptive, technology slip through the cracks, implying that Magna could easily afford the cost of the test. "If we, as Magna, invested in every early stage development somebody brought in, we would lose a lot of money," he explained. "Certainly, to say industry should look at all these is not the answer, because there are just too many out there. The resource drain would be completely unsustainable. Ultimately, when somebody comes up with the next big winner, that person also has to have some kind of spirit or energy to push it forward or it fails. That's how it has worked and that's how it will continue to work." In other words, the ball was in Heins' court.

There was no doubt Heins had that spirit, but he also had a tendency to get quarrelsome. At first introductions, he could be a calm teddy bear, and this often worked to his benefit by disarming those he approached. But his frustration could also get the better of him. In February 2009, he was invited to pitch his technology to a panel of judges as part of a local innovation event in Ottawa, but it ended in a shouting match. One of the judges was multimillionaire Robert Herjavec, who is a judge on the CBC television show *Dragon's Den*, a kind of *American Idol* for new business ideas. Heins didn't get the reaction he wanted, but he also didn't like the way Herjavec flaunted his wealth — his big gas-guzzling cars, mansion, and yacht.[11] It stood for everything Heins was against, and this led to a firework of words in front of an audience of hundreds. It wasn't the best way for an underdog inventor to make friends, yet Heins has never hesitated to question motivations, to taunt, to accuse others of being cowardly for not believing. His preachy explanations for why people don't believe, don't return calls, won't see a demonstration, or won't champion his cause can sound like conspiracy theories in which he is a lone martyr fighting for the noble cause. "Some inventors are very frustrated at not getting heard, and they have a style that's hard to listen to," Amory Lovins, chief scientist at the Rocky Mountain Institute, once told me. "It goes with the territory." That was becoming increasingly evident.

Still, Heins' ability to build new relationships never ceased to amaze. In late 2008, for example, Heins notified me that he had been contacted by Canadian rocker Neil Young. What could a music legend like Neil Young possibly want to do with a struggling inventor in Ottawa? Most fans of Young would know that the 60-something musician is an environmentalist who likes big cars, kind of like being a tree hugger who collects chainsaws. But Young doesn't see it that way. Big is good, in his view, as long as the carbon footprint is small. So the "Godfather of Grunge" (a nickname he earned in the 1990s because of his influence on Seattle bands such as Pearl Jam and Nirvana) launched a project in 2008 to prove a big car could leave behind a small footprint. He would "repower" his 1959 Lincoln Continental convertible so it could run on electricity and be backed up by renewable fuel or compressed natural gas. He wanted to make automotive history by giving his 2.5-ton, 19.5-foot vehicle the ability to travel for several hundred miles using the clean-energy equivalent of just a few gallons of gasoline. Young called the rebuilt vehicle the LincVolt and he planned to film every step of his journey for a documentary production.

As part of this ambitious project, Young also entered the LincVolt into the Progressive Automotive X PRIZE competition, which promised a $10-million award for the best new vehicle design that could achieve 100 miles per gallon. To make good on his vision, Young hired a team of engineers and car tinkerers led by uber-mechanic Johnathan Goodwin, a professional car hacker who has turned gas-guzzling Hummers into 60-mile-per-gallon workhorses that sip French-fry grease and electricity.[12] The LincVolt team, working out of Goodwin's workshop in Wichita, Kansas, included German-born physicist and inventor Uli Kruger and U.S. robotics expert Paul Perrone. "We're doing a lot of different stuff that we don't think has been done before," Young told David Letterman on a *Late Show* appearance on January 30, 2009. "If I fail, who cares? I'm not afraid of failure."

LONG MAY YOU RUN

This pioneering attitude may explain Young's willingness to approach Thane Heins. The rocker was determined to think outside the box, and here was some inventor in Ottawa about as outside the box as one can get. But don't think for a moment that Young's rock-star status would earn him any reverence from Heins, as it didn't take long before the teddy bear inventor showed he could growl. The two men exchanged much information by email, and while Heins invited Young to visit Ottawa and Young offered to fly Heins to Wichita, there is no evidence they ever met. Heins indicated from the start he didn't want to be in Young's film and didn't like the idea of the x prize because he believed that collaboration, not competition, was a better approach. In an April 2008 email, he told Young that his participation in the x prize meant he didn't "get it," to which Young graciously replied, "I respect your conclusion about us, but I think it is hasty." Young later added, "Personally I don't give a rat's ass whether we win the prize or not. We are in a race against time. Not each other. However, it does rally a lot of energy and focus on the task."

In May 2008, the grizzly in Heins resurfaced. As he often did, Heins sent out an email broadcast alerting individuals, including Young, to new test data and demonstration videos uploaded to YouTube. Young, a very private individual, was included openly on the email list and wasn't pleased. He sent a gentle message asking that Heins not include his personal address again, but also assured him that "this in no way negates my enthusiasm and curiosity about your project." Heins lashed back at the rock legend. "Are you serious?" he asked in a follow-up email. "I am sorry but the information I sent and the ramifications of what it can mean to humanity far outweigh your personal email address, not to mention your LincVolt project." Then Heins really let it spill, revealing a delusional side to his character. "What I have done is bigger than Einstein, Tesla, or anything Edison ever did — in fact, it is bigger than almost every previous scientific achievement combined." If that didn't sour the relationship, there was more to come. "I hope your enthusiasm and desire [to work together] does not hinge on the fact that I agree with every shallow thing you say," Heins concluded. Ouch! In a short reply, Young continued his gracious ways. "I am completely over it," he wrote back.[13]

It's little surprise that Young backed off from that point

forward. The two still occasionally exchange email, but Young's position is simple: don't just talk about it, build a machine that puts out more power than it consumes. "Thane is doing the development independently and when it is done, we will attempt to demonstrate the tech in a way that is meaningful to the public," Young wrote me in an email in summer 2010. "We have not seen the results of his work in a real application for everyday use and can not verify them at this point. That is what our aim is to do. It sounds interesting and although fringe, he is very committed and we appreciate his energy for the technology." Young faced a setback himself that November when a fire broke out in the warehouse where he stored his LincVolt, which was partially damaged.[14]

Life in 2010 became more difficult for Heins. In March, he was asked to leave his lab at the University of Ottawa. Professors there, including his supervisor, Professor Habash, now avoid him, and many of the bridges he built during 2008 and 2009 have since crumbled. "We are in a cycle where we demo the technology successfully and then they Google me and it is over," Heins explained to me in an email. "I have been labeled as a nut bar by anyone who could help us and a fanatic because I keep trying. I have been told to cease using Dr. Habash's name, Ottawa University, or Ottawa University electrical engineering department names." Heins is undeterred. He continues to forge new relationships, hold open demonstrations, and reach out to corporate giants such as General Motors, ABB, and Philips on the slight chance he can one day gain a credible ally. He has refined his motor-generator technology and attempted to come up with a backing theory for its behaviour. As of June 2011, however, there was no sign of a working prototype that could drive a real-world application. Within the engineering community, Heins has earned some arms-length encouragement, but so far no one has publicly backed him, likely for fear of repercussions on their own careers.

Is Heins a victim of scientific groupthink? Is he being ostracized by a society unwilling to believe? Certainly, there is a large portion of people out there who are outright dismissive of Heins strictly because of the extraordinary nature of his claim. Would a reasonable person expect any less? But Heins has done himself no favors by coming off at times as the kind of "wild-eyed" inventor that even sympathetic ears like an Ed Beardsworth (or Neil Young) would rather keep at a distance.

To this day, I have mixed views about it all. One side of me — the practical and sensible side — sees an outlandish assertion by a man with a volatile personality and an unwillingness to admit he may be wrong in the face of what science says is impossible. The other side — the hopeful, naive side with the "will to believe" — can't help but wonder whether Heins has truly stumbled onto some previously untapped source of magnetic energy and that, given time and patience and the right people by his side, he may some day surprise us. This side of me credits Heins as an underdog for having the "boldness of ignorance," a term that Nikola Tesla once used to describe himself.

It's important to keep in mind that Heins is not unique in his specific pursuit. He is one of among hundreds of individuals working away on over unity devices that claim to produce more energy than they consume. Science tells us you can't create or destroy energy, but the still mysterious nature of magnetism has left many questions unanswered. Why else would NASA reach out to an individual like Heins? "There are people who will go on for volumes, literally, about how Maxwell's equations are incomplete and were simplified in the early days to leave out other cases. That if you build a simple and clever device, you can get energy from magnetic fields," said Beardsworth, who looked at several over unity motor ideas as part of his role in the Hub Lab. "So far, nobody has built something [that works] that has been objectively, carefully, and properly measured."

Whether it's Randell Mills at BlackLight Power or Thane Heins at Potential Difference, or anyone else with spirited determination and that boldness of ignorance, I'm prepared to be surprised.

Are you?

Notes:

1 Patrick Doyle, "Rebirth of Cool," *5820 Magazine*. May 2010.

2 Gargi Chakrabarty, "Cool Prospects for Green Colo. AC Company," *Denver Post*. May 22, 2009.

3 Erico Guizzo, "Loser: Hot or Not?" *IEEE Spectrum*. January 2009.

4 Erik Baard, "Harvard M.D. Challenges Big Bang Theory," Space.com. May 23, 2000.

5 Chris Morrison, "BlackLight Power Bolsters its Impossible Claims of a New Renewable Energy Source," NYTimes.com. October 21, 2008. Also available at http://venturebeat.com/2008/10/21/blacklight-power-bolsters-its-impossible-claims-of-a-new-renewable-energy-source/.

6 "BlackLight Power Inc. Announces Production of Electricity from a New Form of Hydrogen," press release, BlackLight Power. November 29, 2010.

7 Michael Kanellos, "A Gigawatt from a Liter of Water?" GreentechMedia.com. November 29, 2010.

8 Michael Stroh, "Turning Water Into Fuel," PopSci.com. November 13, 2007.

9 John Herrman, "Independent Jury Rules on Steorn's Free Energy Machine: Guilty (Of Not Working)," Gizmodo.com. June 23, 2009.

10 Natan Weissman, "Thane Heins Perepiteia Explained," SciScoop.com. May 21, 2009.

11 Tyler Hamilton, "Electric Motor Polarizes Opinion," *Toronto Star*. February 28, 2009.

12 Clive Thompson, "Motorhead Messiah," *Fast Company*. November 1, 2007.

13 Heins copied me on many of his email exchanges, including those with Neil Young.

14 Jonathan Schultz, "Neil Young's Hybrid LincVolt Survives Fire," Wheels.Blogs. NYTimes.com. November 16, 2010.

CONCLUSION
Reasons to Hope

"I'm not a doomsayer. Everybody knows
I'm not. But we're facing deep problems
and we need everybody's help."
— Stanford Ovshinsky, prolific American inventor

No matter how compelling an energy breakthrough, it has been argued that the nature of energy transitions will prevent such a breakthrough from having an immediate effect on our lives. Energy expert Vaclav Smil, a professor of environment at the University of Manitoba and one of the most prolific writers on the subject, likes to routinely remind us of this reality. "Historical evidence shows that energy systems, the most complex and capital-intensive mass-scale infrastructures of modern societies, are inherently inertial, and that our determination can accelerate their change but cannot fundamentally alter the gradual nature of their evolution."[1] To think otherwise is naive and to suggest that a potentially disruptive energy technology can have a substantial and speedy impact is "irresponsibly exaggerating the promise," Smil writes. "They should become an important part of an energy solution, but they cannot make as much difference, and as fast, as is now so commonly believed."[2]

This must always be kept in mind, as any new innovation — incremental or disruptive — cannot escape going through the phases of discovery, development, demonstration, commercialization, and maturation, and there are dozens of hurdles along the journey toward mainstream acceptance. I have attempted throughout this book to identify some of these hurdles and to explain how good technology on its own isn't enough to guarantee success. It's only one step in a long journey.

Putting hundreds of massive solar power stations into orbit may be technically possible, and it may prove to be economical, but is it practical when weighed against less risky clean-energy alternatives? Can a company such as Solaren raise the enormous amount of financial capital required to do it? Can Gary Spirnak satisfy the public's concerns around safety, and will he be able to prove that taking solar to space is better than embracing its full potential here on Mother Earth?

Louis Michaud knows he can economically harness waste heat and ocean heat to create man-made tornadoes that generate power, but can he raise enough money to demonstrate it? If he does, can he overcome the perception, however false it may be, that his unorthodox idea is too dangerous to pursue on a large scale?

Will Paul Woods at Algenol overcome many of the same perceptions, or will established fossil fuel interests and industries get in his way and forever keep him a niche biofuel player? Can his unique process deliver ethanol at the cost and quantities that matter?

Can Michel Laberge convince venture capitalists, government agencies, and regulators that General Fusion can do what multibillion-dollar international consortia cannot? Can he find people with the right skills to help him achieve his goal, and will he have the staying power to make it to a finish line that the nuclear industry, dominated by fission reactor technologists, doesn't want him to pass?

Even if EEStor has a game-changing energy storage technology that's proven to work, will it be hindered by the stubborn personality of its founder and be marginalized by competitors — old and new — who see the company as a threat to their own, incremental innovations? Will a technology that wows under controlled conditions be revealed as fragile and unreliable when mass-produced for real-world applications?

One, some, or all of these companies may overcome one, some, or all of the barriers that lie ahead, and they may even achieve commercialization after many years of trying. But even if commercialized, the path to maturation can take decades more because of the slow-moving nature of existing energy systems — a complex and tightly bound interaction of machines, wires, pipelines, fuels, energy suppliers, and energy consumers that, taken together, are deeply resistant to change. At the same time, most individual stakeholders in this old boys' energy club have an interest in stonewalling change to assure their own survival, and behind them is a financial community that has grown fat by avoiding risk and not rocking the boat. We cannot underestimate the grip that fossil fuels have on us or the challenge of transitioning to a low-carbon economy.

The venture capital community is one group that has learned this through experience. Not so long ago, Silicon Valley and its tech-savvy ilk got the green bug, and many of the multimillionaires and multibillionaires who made it big during the computing, dot-com, and telecom booms decided that the next big opportunity was in clean energy and green technologies. They raised many billions of dollars and started pumping this money into next-generation solar technologies, biofuels, batteries, fuel cells, vehicles, smart grid equipment, carbon-capture systems, geothermal projects, and other "greentech" innovations. Between 2001 and 2008, for example, global venture capital investment in clean technologies jumped from $507 million to well over $8 billion.[3] The recession caused a dip in 2009, but investment began to climb again in 2010.

What these investors realized soon enough is that, while the opportunities were massive, breaking into the energy business takes much longer and requires far more capital than, say, building the next big e-commerce site or search engine, designing a must-have software application, or even making the next hot consumer electronics gadget. Rapid change may be built into the DNA of the information technology sector and the Internet, but the energy sector is a beast dominated by regional monopolies and global oligopolies. New innovation is embraced slowly, and far too often it's viewed as a nuisance or threat rather than an opportunity. The energy sector has its own pace, set by the multi-trillion-dollar global infrastructure it is determined to protect. This reality can be discouraging, even for the world's richest man. "You're not going to have a lot of people putting down money when . . . the time period to get something done is actually longer than the length of the patent," Microsoft co-founder Bill Gates lamented during an energy conference in 2010. "People like to work on things that happen during their lifetime."[4] Silicon Valley, with all its influence, money, and business savvy has been forced to do something it doesn't like to do: be patient.

Venture capitalists typically want in and out of investments in less than five years — preferably within two or three years — and many of the most promising new energy technologies can't be fully developed and deployed under such a timeline. "It might be fine with software, but not with innovation around biomimicry," said Jay Harman of PAX Scientific. This was a common observation with the inventors and entrepreneurs I interviewed for this book, and Michael Brown, the venture capitalist behind General Fusion, said it represents a serious barrier to innovation in the energy sector. "When Silicon Valley tries to play in clean technology hardware, it gets its head served on a platter," said Brown, adding that the investment philosophy that thrived during the high-tech boom simply doesn't work with energy. "The model is broken."

This and other barriers are delaying the essential and inevitable decline of fossil fuels. In the electricity sector, for example, the International Energy Agency estimated that coal, natural gas, and other fossil fuels used to generate electricity will still be the dominant energy source in 2035 and that it will take 25 years for their share to drop from 68 percent to 55 percent of the energy mix. Solar power, by comparison, will only grow to represent 2 percent in that timeframe. Renewables — the total mix of wind, hydro, solar, and others — could reach one third of electricity supply, up from 19 percent in 2008. But to get there, the agency said $5.7 trillion would need to be invested, and many believe that still won't be enough to avert the worst effects of climate change.[5]

From this perspective, the glass is half empty.

THERE IS AN UPSIDE

I'm a glass half full type of person, so let me offer a more encouraging perspective. There is a general recognition out there that *we are* in a period of meaningful transition, and that investment is increasingly shifting to low-carbon energy sources and technologies. There is also no sign that this trend is going to fade, or that green energy is some kind of fad. In 2008, for example, an important milestone was hit: more dollars were spent adding renewable-power capacity to the world's electricity systems than on fossil fuel and nuclear energy sources combined.[6] Renewables, to be more precise, captured 56 percent of the $250 billion spent on new power capacity that year. This remained true in Europe and the United States in 2009, and while there will be spurts and sputters along the way, the trend is expected to hold steady over the coming years.[7]

The global Boston Consulting Group (BCG) isn't exaggerating when it reports that there has been an "unprecedented explosion of interest" in alternative energy technologies during the first decade of the 21st century. Indeed, the consultancy singled out advanced biofuels and concentrated solar power as two technologies in a position to "disrupt the status quo" *without* the crutch of subsidies by 2025. Could these, to borrow a term popularized by author Malcolm Gladwell, be tipping points that contribute to more rapid change in the energy sector? As BCG pointed out, "Even in the relatively slow-moving energy space, there are cautionary examples of how quickly fundamental assumptions can be overturned."[8] Advanced drilling technologies, to name one recent example, enabled the economic collection of shale gas and changed the fortunes of the natural gas sector in a matter of years.

I want to emphasize the use of the word *unprecedented* above, not to question the historical fact that energy transitions are inherently inertial, but to suggest that the energy transition currently washing over us cannot be directly compared to transitions of the past. There *is* something different happening this time around and much of it is not exaggeration or wishful thinking. It's real, and it has momentum. When I started reporting on the energy sector in 2005, I was one of just a handful of reporters and bloggers who focused on clean energy and green technologies, and one of the first in North America to have a column in a large daily newspaper that was dedicated to the subject. Today, I am but one small voice in a sea of dedicated news sites, columns, blogs, Facebook pages, and Twitterers all covering different angles of this clean energy revolution and advocating for a faster transition away from fossil fuels. We may complain that the transition is going too slowly — it can never move fast enough — but looking back it's amazing we have come this far so quickly.

Across North America, municipalities, states, and provinces are tightening their building codes, promoting energy efficiency, mandating the use of renewable energy, supporting the deployment of electric vehicles, and reinventing their electricity distribution and transmission systems as smart grids. More than half the states in America now mandate a certain percentage of renewable energy in their electricity mix, while my home province of Ontario, which aims to go coal-free by 2014, has introduced a Green Energy Act and a feed-in-tariff program that is expected to add thousands of megawatts of renewable energy capacity to the grid over the coming years. All of this, again, is unprecedented.

Also unprecedented is that most major automakers on the planet — American, Japanese, Chinese, Korean, and European — are coming out with electric cars; General Electric has committed to purchasing 25,000 electric vehicles within five years for its own fleet and its customers' fleets; electric vehicle charging stations are starting to sprout up at university and corporate campuses, hotel parking lots, big-box retailers, and airports and are increasingly operated by a new breed of company selling electrons as a service the same way mobile phone companies sell us minutes on their networks. Something different is clearly happening here.

There is profound change also going on with the U.S. military, America's largest single energy consumer, which must provide electricity to hundreds of bases throughout the world and fuel to hundreds of thousands of vehicles, aircraft, and ocean-going vessels. The Department of Defense consumes 300,000 barrels of oil a day, reflecting what Admiral Mike Mullen, chairman of the joint chiefs of staff, described as a "burn it if you've got it" mentality once held by the army, air force, and navy. "We just held a very conventional view that fuel was cheap, easy, and available, without ever really connecting it to any broader geopolitical implications," Mullen told guests at an energy security conference in fall 2010. "Clearly, that is not the world we're living in today."[9] The admiral's comments were sobering. He said climate change, while it will create a humanitarian crisis, also creates conditions of hopelessness that can spark more world conflict. On top of that, the cost of getting oil to the battlefield is rising and putting soldiers at greater risk.

It's why the U.S. Air Force is testing renewable jet fuel on its A-10C Thunderbolt jets and wants half of its fleet of planes to run on a blend of biofuels by 2016. The navy, meanwhile, wants to cut its use of non-tactical petroleum in half by 2015 and has committed by 2016 to launching a "Great Green Fleet" of battle-ready ships, submarines, and planes that would be fueled entirely by biofuels. Navy Secretary Ray Mabus, who once served as ambassador to Saudi Arabia, wants renewable energy to represent half of all power and fuel needs of both the navy and marines by 2020.[10]

Solar technologies are also expected to play a major role on the ground to reduce the carbon "bootprint" of military personnel. The army, for example, is building a 500-megawatt solar thermal and photovoltaics plant in the Mojave Desert to power its Fort Irwin base in California.[11] There is no reason to believe this is just good public relations as part of attempts to clean up the image of a U.S. war machine, as Admiral Mullen made clear. "This effort is not merely altruistic. It is essential," he said. "Failing to secure, develop, and employ new sources of energy or improving how we use legacy energy systems creates a strategic vulnerability and, if left unaddressed, could threaten national security."[12] This kind of talk is unprecedented, as are the Pentagon's commitments to lead the transition to low-carbon energy sources.

If all of this is encouraging, it pales when measured against the actions of the Chinese, which according to the chief economist of the International Energy Agency are ushering in a "new age in the history of energy."[13] The world's most populated, fastest-growing, and now most energy-consuming nation is an easy target for Conservative politicians in the developed countries of the West, particularly in the United States. Why, they argue, should America commit through international treaty to reduce its carbon emissions when China has yet to agree to mandatory reductions of its own? That argument is a red herring. China represents for many an excuse to do nothing, but the mistake here is to assume that China, by not yet agreeing to mandatory reductions, is actually doing nothing. Far from it, China recognizes that its own growth is making it dangerously dependent on imports of coal and oil, and that its demand for such fossil fuels is driving global prices higher. This situation, combined with its concern for local pollution and climate change, has led China to aggressively embrace renewable energy and clean technologies to the extent that it now represents an imposing economic threat to the United States.

In 2009, China surpassed the United States as a top investor in green technologies and has quickly emerged as the world's clean energy powerhouse.[14] It is now the world's top manufacturer of both solar and wind technologies, a position it secured in a matter of years. China's 12th Five-Year economic plan aims to retain that dominance with ambitious domestic targets for renewable-energy and "green" vehicle development. Population and economic growth means that coal-fired power generation will continue to grow, but as a percentage of overall supply it will drop to 63 percent from 72 percent as more renewable sources are added to the mix. China expects to nearly quadruple its wind power capacity by 2020, though this may be a conservative estimate. Some experts say a six-fold increase to more than 240 gigawatts of capacity is doable. Solar PV installations are expected to increase 20-fold over the same period to more than 20 gigawatts, an outcome that would significantly drive down the global cost of solar technologies. China is also going big with electric vehicles. By 2015 it expects to have a network of 4,000 EV charging stations, rising to 10,000 stations by the end of the decade. As well, it's widely expected that Chinese automakers and their technology partners will play a lead role in driving down the cost of vehicle battery technologies. This would leave the United States, based on its current trajectory of investment and deployment, in its dust.[15]

It's no wonder U.S. energy secretary Steven Chu talks about America's "Sputnik moment." The energy race is on whether or not politicians and citizens realize it. Chu realizes it. Sooner rather than later, America's survival instinct must kick into higher gear. Once again, this is all unprecedented. History cannot explain with accuracy how this future will unfold. But the general direction is increasingly obvious. "In the next 30 years, we'll be in the biggest industrial reformation we've seen in modern times," said Nicholas Parker, co-founder and executive chairman of the Cleantech Group, an industry research and consulting group on the front lines of this reformation. "What we're seeing evidence of around the world is that certain jurisdictions get that this is the new economic space race."

What I've discussed above partly explains why I believe the transition away from fossil fuels could be faster than energy transitions of the past. This isn't about a single form of clean energy taking on entrenched interests behind an established global infrastructure of unparalleled scale and influence. This isn't about wind versus coal, biofuels versus oil, electricity versus gasoline, or solar versus natural gas. This is about unprecedented innovation, technological diversity, global competition, information flow, public awareness, environmental degradation, resource depeletion, government policy, military initiative, and concern for national and global security — all of it chiseling away at the foundation of our carbon-based economy. This is about sand-blasting a wall. This is about a collective consciousness and momentum that can make giants, no matter how big, fall.

BLACK SWANS

I may be wrong, of course. We've always done a poor job of predicting the future where technologies are concerned, and while I believe the general direction of our energy transition is right, the pace of change could very well prove disappointing. Forecasting truly is a mug's game. Outcomes rarely materialize as predicted, and on the opposite extreme, some events unexpectedly emerge that bring about rapid change and surprising outcomes. Author and professor Nassim Nicholas Taleb calls these unexpected events "black swans," and they can blindside the optimists and the pessimists alike. "Many events have taken place and new technologies have appeared that lay outside the forecasters' imaginations," Taleb wrote. "Many more that were expected to take place or appear did not do so."[16]

I am fascinated by this idea of black swans, and how unexpected discoveries or unforeseen innovations can have such dramatic impacts. Nikola Tesla was a black swan. The future well-being of humanity may depend on one of these rare events, as incremental steps can point us in the right direction but might not ultimately get the job done. This book, in many ways, is about the potential for black swans to emerge from the many thousands of entrepreneurs, inventors, scientists, and engineers dedicating their lives to solving an energy crisis that is clear and present.

It's about how the possible can unexpectedly emerge from the impossible, however improbable. Perhaps there is a black cygnet within these chapters? "We build toys," Taleb wrote. "Some of those toys change the world."[17]

It follows that the more toys we build and explore, the greater the chance of a black swan emerging. The fact that so much invention and innovation these days is focused on tapping into clean, economical, abundant, and sustainable forms of energy can only improve our odds that disruptive change, should it come, is the change we desire and need. "I feel that right now in the world there's a ton of bright, knowledgeable physicists, scientists, and engineers who are pushing toward solving these critical energy problems, and I truly think that is going to save humanity," said John Paul Morgan, founder of Toronto-based Morgan Solar and inventor of a concentrated solar cell technology that can compete head to head with coal power. I met Morgan for dinner one evening and, over a beer or two, we chatted somewhat philosophically about the need to take risks, to explore new territory, to remain open minded, but also to aim for positive change. "If I, and thousands of other people I don't know and will never meet, all push in this direction of innovating around clean energy, some of us will be successful and the world will be better for it," he said.

It truly is a movement. I don't know about you, but I'm tired of all the doom and gloom. I want something to cheer for, and I want hope for the future of my children, and their children, and so on. This book is meant to inspire, give hope, and offer comfort that there is much to cheer for — that humanity isn't resigned to choking on its own industrial past. Isn't it incredible and heartening that a small group of passionate individuals in a suburb of Vancouver are trying to crack the nuclear-fusion nut? That a retired mechanical engineer wants to harness energy from man-made tornadoes? That a former jet-fighter pilot is trying to make long-range electric cars a commercial reality? That a seasoned aerospace executive has the audacious vision of collecting solar energy in space and beaming it back to Earth? Or that there are individuals out there daring to question some laws of physics if it means bettering our life on this planet?

We need these risk-takers and visionaries so willing to expand the realm of possibility. Not only are they part of that collective consciousness that can make giants fall, they are the potential black swans who can carry us to the low-carbon world we need, sooner rather than later. They are the unappreciated warriors on the front lines of necessary change, and whether they succeed or fail, we owe them our gratitude. That's because they succeed by making the attempt — by being mad like Tesla and building the "toys" that expose us to unexpected, disruptive change. And while society's skepticism and ridicule may be harsh and biting, it is a crucial condition of their journey. For Nikola Tesla, it was the bittersweet fruit of invention, a point he made clear in a 1913 letter to financier J.P. Morgan:

[It is better] in this present world of ours that a revolutionary idea or invention, instead of being helped and patted, be hampered and ill-treated in its adolescence — by want of means, by selfish interest, pedantry, stupidity, and ignorance; that it be attacked and stifled, that it pass through bitter trials and tribulations, through the strife of commercial existence. So do we see our light. So all that was great in the past was ridiculed, condemned, combated, suppressed — only to emerge all the more powerfully, all the more triumphantly from the struggle.

Notes:

1 Vaclav Smil, *Energy Myths and Realities: Bringing Science to Energy Policy* (Washington, D.C.: AEI Press, 2010), 59.

2 Vaclav Smil.

3 "Clean Technology Venture Investment Reaches Record $8.4 billion in 2008 Despite Credit Crisis and Broadening Recession," press release, Cleantech Group. January 6, 2009.

4 Tyler Hamilton, "Overhauling Energy Will Be Slow, and Expensive," *Toronto Star.* August 16, 2010.

5 "World Energy Outlook 2010," International Energy Agency. November 2010.

6 James Kanter, "Clean Energy Funding Trumps Fossil Fuels," Green.Blogs.NYTimes
    .com. June 3, 2009.

7 Just because the world invested more in renewables than fossil fuels in a given year
    does not mean more megawatts (generation capacity) or megawatt-hours (electricity
    supply) of renewable energy were added compared to fossil fuels. Those are two
    other important milestones we have yet to meet.

8 Balu Balagopal, Petros Paranikas, and Justin Rose, "What's Next For Alternative
    Energy," The Boston Consulting Group. November 2010.

9 Admiral Michael Mullen, speech delivered at the Energy Security Forum. October 13,
    2010. Mullen is the chairman of the Joint Chiefs of Staff in Washington, D.C.

10 "Remarks by Honorable Ray Mabus, Secretary of Navy," speech delivered at the
    Pentagon. January 21, 2010.

11 C. Todd Lopez, "Army on Track to Power Fort Irwin With Sunshine," U.S. Army
    News Service. August 7, 2009.

12 Admiral Michael Mullen.

13 Spencer Swartz and Shai Oster, "China Tops U.S. in Energy Use," *Wall Street
    Journal*. July 18, 2010.

14 "Who's Winning the Clean Energy Race," report by The Pew Charitable Trusts.
    March 2010.

15 The Climate Group, "Delivering Low-Carbon Growth: A Guide to China's 12th
    Five Year Plan," March 2011. TheClimateGroup.com. (Study commissioned by the
    HSBC Climate Change Centre of Excellence.)

16 Nassim Nicholas Taleb, *The Black Swan* (New York: Random House, 2010), 162.

17 Nassim Nicholas Taleb, 170.

ACKNOWLEDGEMENTS

Shortly after I came up with idea for this book, I began shopping it around. It was a lengthy process, but I was fortunate to have so many publishers in Toronto open their doors. ECW Press, located just a short jog from my home in Toronto's Beaches neighbourhood, turned out an ideal fit in so many ways. I thank co-publisher Jack David for buying into my vision and managing editor Crissy Boylan for what felt like, from my perspective at least, a painless editing process. My gratitude extends to the rest of ECW's team, including Simon Ware, Erin Creasey, and Jennifer Knoch. It was a pleasure working with you all.

Whether they knew it or not, many individuals played an important role as this project unfolded. First and foremost, a big thank you to my friend Greg Kiessling, who despite being an ocean away read each chapter as I finished it and offered invaluable feedback and advice. Greg also opened up his professional network to help me spread the word about *Mad Like Tesla*. On a subconscious level, I'm sure our past chats over coffee played an early role in inspiring parts of this project. I am also indebted to Jim Balsillie for keeping me focused. Jim took the time out of his immensely busy schedule to read early chapters and share his thoughts.

Similarly, thank you to Toby Heaps for never turning down an invitation to brainstorm over a beer. Not only did Toby introduce me to ECW, but our regular discussions on energy, sustainability, and green politics helped me to refine my ideas and approach to this project. I must also acknowledge my friend Jose Etcheverry for both his personal and professional support, including sticking out his neck to help me when my own efforts failed.

There are so many others who deserve my thanks for contributing directly or indirectly to this project. They include Nicholas Parker, who first inspired me to start writing about clean technologies; George Monbiot, who unknowingly convinced me to leave my day job and pursue this project as part of a freelance career in our first meeting; and the *Toronto Star*, for embracing my Clean Break column and giving me the platform on which to write about so many inspiring stories in the energy and technology sectors. Alfred Holden, the best editor I've ever worked with at the paper (or anywhere, for that matter), not only shared my enthusiasm for these stories but fought for the real estate I needed to tell them properly.

It should be emphasized that this book would be empty were it not for the "underdog inventors" and entrepreneurs profiled within its pages. Most opened up their doors and their lives, letting me get close enough to do the job I needed to do. My deepest gratitude goes to Michel Laberge, Doug Richardson, Gary Spirnak, Cal Boerman, Louis Michaud, Jay Harman, Paul Woods, Ian Clifford, and Thane Heins. Many thanks as well to Tom Rand, Vinod Khosla, Eric Mennell, the Morgan brothers, Janine Benyus, Amory Lovins, Rick Whittaker, Vicky Sharpe, Michael Brown, Peter and Eva Glaser, Kerry Emanuel, Michael Blieden, Ed Beardsworth, Dave Pascoe, Neil Young, and Dave Toms — not to mention many loyal readers of my Clean Break blog.

And, of course, recognition must go to the man who inspired it all: Nikola Tesla.

Closer to home, I offer a heartfelt thanks to my family and friends for their unwavering support. To my parents, Jo-Ann King and Paul Hamilton, and my in-laws, Bill and Denise McMurchie, for your love and encouragement; to my sister, Tracey Boriska, for helping me stay sane when anxiety occasionally set in; and to my brother-in-law, Neil McMurchie, for being my engineer-on-demand and helping me understand often complex technical issues before my brain exploded. The Boriskas, Constantinos, Macs, Eakins, and Turners — I wish I could have spent more time with all of you over the past 18 months.

Finally, I could not have done this without the support of my dear wife, Lyne McMurchie, and my two smart and heart-achingly beautiful girls, Claire and Ruby. Lyne was not only my unofficial editor; she was my sounding board, my voice of reason, and my bringer of late-night tea. At the same time, she tolerated my downs, kept all in order when I needed to dash off for trips, and got the girls bathed and off to bed many nights while I typed away oblivious to the time. Thank you for your tolerance and love, and for indulging me in this pursuit.

As for my little bunnies, as curious as you are, you have had to suffer through my ramblings about clean energy, climate change, and ways to lower our collective impact on this planet. Even if you were just pretending, thank you for showing interest in what your old man had to say. My gut, however, tells me you weren't pretending, and that delights me to no end.

INDEX